The Heart of Our Earth

Praise for this book

'Mining's curse may not be new to Latin America, but its dirty paw marks now tread larger than ever – and are expanding by the day. *The Heart of Our Earth* lifts the lid on the geopolitical machinations and economic interests that lie behind the continent's winner-takes-all obsession with destructive extractivism. Drawing on prodigious, first-hand research, this fiercely argued account unpicks the myth of 'sustainable mining' and brings fresh light – and hope – to the cause of anti-mining campaigners. Tom Gatehouse does not seek to paint a romantic picture of a post-extractivist world, free of mined metals and minerals. Instead, he brings a much-needed and too often overlooked perspective to the narrative of commodity-led growth; one that puts people – not profits – squarely at its centre. An urgent, necessary, and timely book.'

Dr Oliver Balch, freelance writer and PhD in Latin American Studies from the University of Cambridge

'In *The Heart of Our Earth*, Gatehouse does what he does best: highlight the voices and power of movements making change. Focused on the impacts of the Latin American "mining boom" over the past few decades, the book artfully weaves testimony with action to highlight not only how frontline communities are struggling against predatory mining, but how these communities are winning.

Moving from Mexico to Argentina, Gatehouse explores some of the most important cases of Latin American "people power" in the Americas. We hear from frontline activists, NGOs and scholars as they interpret their struggles and their realities from their direct experience with the mining industry in its various stages.

Gatehouse's reporting doesn't stop with community voices and part of the book's strength is how he successfully keeps an eye on the corporations (most of them Canadian), and governments that seek to benefit from the violence perpetuated against communities and their environments. This "new-imperialism" has its roots in neoliberal institutions and legal regimes which prioritize profit maximization at all costs. Their active implementation has enabled some of the largest and most toxic mines history has known in order to satisfy global demand for an ever-expanding economy of things.

As we face impending collapse due to our ecological crisis and warming planet, *The Heart of Our Earth* contains a sombre warning about the potential for green-capitalist solutions: unfettered demand driven by private ownership and capitalist accumulation require dispossession. Despite mining corporations' best efforts to greenwash their operations, they can't escape their material realities. As Joan Kuyek poignantly noted, mining is a waste management industry that makes its profit on the mining of finite resources. Sustainable/Green mining isn't just a misnomer, it's an active attempt to generate consent for further dispossession. As evidenced in *The Heart of Our Earth*, frontline communities in the Americas aren't demanding better conditions for an industry where they have reaped no benefits and borne all costs. What they are demanding is something radically different, wherein no one, no community and no collective environment, is sacrificed in the name of capitalist accumulation.

An essential read for anyone interested in contemporary mine struggles in Latin America.'

Dr Kirsten Francescone, Assistant Professor in International Development Studies at Trent University and former Latin America Program Coordinator at MiningWatch Canada.

'Unrelenting global demand for minerals is reaffirming Latin America's historically defined position in the world economy as provider of natural resources. With case histories from across the region, *The Heart of Our Earth* provides an up-to-date and lucid assessment of the skewed form of development that this generates. Communities affected by mining are left to challenge a system that serves to enhance local, regional and global inequalities. This they are doing, helping to shift the terms of debate around mining. The book is a parable for our times.'

Dr John Crabtree, Research Associate at the Latin American Centre at the University of Oxford

'*The Heart of Our Earth* tells an insightful story of resistance to large-scale mining across Latin America, skilfully weaving together interviews, case studies, and extensive research, to provide a detailed and accessible discussion of the myriad ways in which the industry is impacting communities, livelihoods, and the environment across the region. Reflecting LAB's long-standing commitment to human rights and social justice, *The Heart of Our Earth* makes an important contribution to shedding light on patterns of inequitable, and often violent, relations between governments, local communities, and corporations, whilst also considering the possibilities for more just and sustainable futures.'

Dr Katy Jenkins, Professor of International Development and Co-Director of the Centre for International Development at Northumbria University

'From the Andean altiplano to the Amazon rainforest, down to the Patagonia region, Tom Gatehouse's new book describes and analyses how indigenous peoples, Afro-descendant communities, campesinos, and urban neighbours debate, fight and resist mining projects throughout Latin America. Based on numerous interviews and cases, Tom presents a portrait of social contestation against mining, respecting the region's social, political and cultural differences. If, as the Canadian educator Judith Marshall says, "a mine, is a mine is a mine," resistance is inventive, multifaceted and varied. Tom's book is a crucial contribution to better understanding this diversity.'

Bruno Milanez, Associate Professor at the Federal University of Juiz de Fora, Minas Gerais, Brazil, and Coordinator of the Politics, Economy, Mining, Environment and Society (PoEMAS) Research Group

'Written in an accessible narrative and solidly grounded in abundant sources and interviews, *The Heart of Our Earth* provides a rich panoramic view of large-scale mining extraction in Latin America and a detailed account of some of the most critical mining disputes from Central America to the Southern Cone in the recent decades.

This book is an incredibly useful resource for specialists, journalists, activists, students, and anyone interested in the region's contemporary history through the lens of mining conflicts and their long-lasting legacies.

In times of increasing violence against environmental defenders and a business-as-usual logic in mining investments, Tom Gatehouse provides a timely and inspiring assemblage of accounts of resistance, legal victories, and struggles for environmental justice and democracy across the region.'

Sebastián Rubiano-Galvis, Postdoctoral Fellow,
International Studies Department, University of San Francisco, USA

The Heart of Our Earth

Tom Gatehouse

Published by Practical Action Publishing Ltd
and Latin America Bureau

Practical Action Publishing Ltd
The Robbins Building, 25 Albert Street,
Rugby CV21 2SD, UK
www.practicalactionpublishing.com

Latin America Bureau (Research & Action) Ltd
Enfield House, Castle Street, Clun, Shropshire, SY7 8JU, UK
www.lab.org.uk

ISBN 978-1-90901-414-5 Paperback
ISBN 978-1-90901-416-9 Hardback
ISBN 978-1-90901-415-2 Electronic book

Gatehouse, T., (2023) *The Heart of Our Earth*, Rugby, UK: Practical Action
Publishing and Latin America Bureau, http://doi.org/10.3362/9781909014152.

Since 1974, Practical Action Publishing has published and disseminated books
and information in support of international development work throughout
the world. Practical Action Publishing is a trading name of Practical Action
Publishing Ltd (Company Reg. No. 1159018), the wholly owned publishing
company of Practical Action. Practical Action Publishing trades only in support
of its parent charity objectives and any profits are covenanted back to Practical
Action (Charity Reg. No. 247257,
Group VAT Registration No. 880 9924 76).

Latin America Bureau (Research and Action) Limited is a UK registered charity
(no. 1113039). Since 1977 LAB has been publishing books, news, analysis and
information about Latin America, reporting consistently from the perspective
of the region's poor, oppressed or marginalized communities, and social
movements. In 2015 LAB entered into a publishing partnership with Practical
Action Publishing.

Cover photo shows four young protesters at an anti-mining demonstration in
Puerto Madryn, Chubut, Argentina, on 26 November 2020. Credit: Alex Dukal
Typeset by vPrompt eServices, India

Contents

Acknowledgements

It is more than five years since LAB decided to commission a book on mining in Latin America. As the project evolved, dozens of individuals and organizations have lent generous support.

Emma Banks, Becky Branford, Matt Kennard, and James Thackara all contributed vital interview material to this book (more details are given in the Preface). Thanks to Matt and Becky for sharing photos from their research trips, and to Matt for sharing his work on *No Bonanza*.

Theo Bradford, Will Huddleston, Elizabeth Pillares, Rowan Ritchie, Matty Rose, James Scorer, and Chris Whitehouse all provided support with transcription and translation.

A special thank you both to David Lehmann for all his encouragement and careful reading of the draft chapters, and to copy editor Clare Tawney for her diligent work ensuring style and consistency throughout the text.

LAB volunteers Jasmine Haniff and Kinga Harasim produced excellent original mining-related articles for the LAB website, some of which have fed into this book.

Generous funding for research and writing came from Christian Aid, the Lipman-Miliband Trust, and David Lehmann. LAB would also like to thank everyone who contributed to the crowdfunding campaign in 2020, particularly Sophie M., and Emily Ryan and Bill Flinn.

Special thanks to Francisco Elías Prada and Angela Rodríguez Torres (Ojos Ilegales Red), who edited the video for the crowdfunding campaign; to Eduardo Vidal, who shot original footage, and to Tatiana Garavito, who voiced the appeal.

We were determined from the outset that our project, like the issues it describes, would not finish with the publication of this book. A generous grant from the Network for Social Change will ensure that we can promote the book, organize discussion of the issues it raises, and chronicle ongoing events and conflicts on *The Heart of Our Earth* website (https://lab.org.uk/the-heart-of-our-earth/). A special thanks also to Mark Brown.

Thanks to War on Want, particularly Sebastián Muñoz and Benjamin Hitchcock Auciello; to Matti Kohonen of Christian Aid; and to Jen Moore from the Institute for Policy Studies (formerly of MiningWatch Canada), all of whom provided support for the project, particularly during the early stages.

I would like to say a special thank you to everyone at London Mining Network, particularly Terry Blackman, Andrew Hickman, Lydia James, Javiera Martínez, and Richard Solly. They provided guidance and contacts, set up interviews, promoted the project, and were indispensable and loyal supporters throughout.

Many other individuals provided contacts, set up interviews, and helped in other ways to make this project a reality, including Letícia Aleixo, Oliver Balch, Paola Bayle, Inge Boudewijn, Phil Chamberlain, Ann Chaplin, Luis Manuel Claps, Jeremias Goransky and family, Valeria Guarneros-Meza, Antonio Ioris, Katy Jenkins, Julia Mello Neiva, Isabela Ponce Ycaza, Ana Reyes-Hurt, Alan Septoff, Waldo Soto, Andrés Tapia, Bruce Wilson, María Fernanda Wray, and Saúl Zeballos.

Several LAB editors and council members provided valuable support: Sue Branford helped to advance the project during early discussions with our partners and provided input throughout. Grace Livingstone helped me with the description of the Atacama Desert in Chapter 7 and contributed photos. Shafik Meghji provided tips on how to promote the book and ensure the continuity of the project. Ainhoa Montoya gave feedback on an early plan for the book and helped unravel legal complexities in Honduras for Chapter 4. David Treece provided contacts from Brazil. Finally, Rebecca Wilson did outstanding work on fundraising and promotion, especially on the crowd-funding campaign in late 2020.

Others at LAB who contributed include Natasha Tinsley, who did important work promoting the project on LAB's social media channels; and Emily Gregg, Tom Kissock, Karoline Pelikan, and all those involved with the Voices of Latin America Indigenous Peoples and the Rights of Nature webinar, which focused partly on the impact of Cerrejón and featured Marcos Brito Uriana (see Chapter 4).

Thanks also to the Friends of LAB patrons for their support: Jon Barnes, Alistair Clark, Malcolm Coad, Paul Garner, Ivette Hernandez, David Lehmann, Elizabeth Lethbridge, Mandy Macdonald, Sophie M., Nick Parker, David Raby, Alison Ribeiro de Menezes, Bert Schouwenburg, Rachel Sieder, Lewis Spence, Pat Stocker, Nick Terdre, and David Treece.

Our partners, Practical Action Publishing, provide all the skills for production and distribution without which LAB could not continue to publish books. Our thanks to Chloe Callan-Foster, Rosanna Denning, Andrea Johnson, Katarzyna Markowska, and Jutta Mackwell.

A very special thanks to LAB editor Mike Gatehouse, for his unshakeable enthusiasm for and belief in this project, and to Verónica Menargues Soriano, for her patience and support throughout. This book would never have made it to publication without them.

Last but not least, thank you to all those who agreed to be interviewed, their families, and communities. I only hope this project can contribute in some way, however small, to greater awareness and understanding of mining and the true costs of the industry for local communities, ecosystems, and our planet as a whole.

About the authors

Tom Gatehouse is a writer, researcher, editor, and translator, who has lived and worked in Argentina, Brazil, and Spain. He worked as editor and project manager on LAB's *Voices of Latin America* (2019) and translated Bernardo Kucinski's novel *The Past is an Imperfect Tense* (2020). His writings and translations have been published on Latin America Geographies, Mongabay, Red Pepper, and Folha de S. Paulo, on the LAB website, and by literary publishers in Portugal and the UK. He lives in Bristol.

Jo Griffin is a freelance journalist and writer who lived and worked in Mexico and Brazil for several years and has continued to report from Latin America. Jo spent ten years on staff with *The Guardian* and her articles have also been published by *The Observer*, *The LA Times*, Al Jazeera, the BBC, and *The Times of India*, among many others. She has worked as an editor and sub-editor and as a reporter for several short films, including *One Man, One City, Three Evictions*, for the Thomson Reuters Foundation, about the history of evictions in Rio de Janeiro. She is currently working on a book about a radical prison system in Brazil. She lives in London with her family.

Preface

From the early 2000s, numerous reports reached Latin America Bureau detailing the rapid expansion of large-scale mining throughout the region. These are documented in more than 180 articles on the LAB website (https://lab.org.uk/category/topics/mining).

In 2016, we decided to commission a dedicated book on the subject and sought funding and support from external partners (see Acknowledgements). At the same time, we were working on LAB's 2019 book *Voices of Latin America: Social movements and the new activism*, which includes a chapter on mining. While doing research for *Voices*, I visited mining regions and affected communities in Argentina, Chile, and Bolivia. Some of these interviews have fed into *The Heart of Our Earth*, while in two cases I have done follow-up interviews with activists whom I first met on that trip.

Generous funding from Christian Aid paid for two research trips in 2018, with one LAB researcher travelling to Chile and Argentina and another to Brazil. The excellent material they gathered has fed into *The Heart of Our Earth*, featuring mostly in Chapters 5 and 7. We had hoped to do further fieldwork in at least Peru, if not also in Colombia and Ecuador, but sadly, the Covid-19 pandemic made this impossible. We redesigned the project in 2020 and launched a successful crowdfunding campaign which enabled us to continue research and writing throughout 2021.

Working during the pandemic posed some considerable challenges. Interviews had to be carried out remotely, while many of the physical descriptions had to be pieced together from articles, videos, and photographs, as well as from conversations with people with first-hand experience of the people and places in question. Inevitably, this is a different book from that we had originally planned, with a broader focus and greater exposition of the social, political, and legal backdrops to these conflicts.

I only hope that we have made up for any lack of local colour with a clear, rigorous, and integrated panorama of the recent advance of mining in Latin America, and the varied contexts in which community resistance has taken root and flourished.

A word on case studies and interviews

The aim when choosing case studies for this book was to be comprehensive but not exhaustive. We have attempted to include a mixture of well-known cases – such as Cerrejón and Yanacocha (Chapter 4) and the tailings dams disasters in Brazil (Chapter 5) – as well as others which have received little or no attention in the Anglophone press. We have attempted to cover as wide a range of countries and scenarios as possible, but with 284 mining conflicts dotted throughout Latin America (according to the latest count from the Latin American Observatory of Mining Conflicts), there are inevitably struggles which either we have been unable to include or to which we have not done full justice.

These include the Fenix and Escobal mines in Guatemala, referred to briefly in Chapter 2; the attempts by Canadian mining companies to mine the Páramo de Santurbán, near Bucaramanga, Colombia; the Arco Minero del Orinoco in Venezuela; various conflicts in Peru, including those around the mines Las Bambas, Tía María, Antamina, and Antapaccay; Canadian junior Belo Sun's proposed Volta Grande goldmine in the Brazilian Amazon; British multinational Anglo America's Minas-Rio iron ore project, also in Brazil; Infinito Gold's Las Crucitas goldmine in Costa Rica; as well as the long conflict in Chile between residents of the town of Caimanes and Minera Los Pelambres.

***

Interviews

Lucrecia Wagner (Introduction, Chapter 6, and Conclusion); Héctor Córdova (Chapter 2); Carina Jofré (Chapter 3); Constanza San Juan Standen, and Stefania Vega (both Chapter 6) were all interviewed in Argentina, Chile, and Bolivia in November and December 2016 by Tom Gatehouse. Carina Jofré was also interviewed remotely in September 2021 by Tom Gatehouse.

Rogelio Ustate Arrogoces (Chapter 4) was interviewed in Colombia in January 2017 by Emma Banks.

Marcos Orellana (Conclusion) was interviewed in Washington D.C. in December 2017 by James Thackara.

Lucio Cuenca (Chapter 2); Sara Plaza and Mariana Cervetto (both Chapter 7); Sergio Cubillos (Chapter 7 and Conclusion); and Cesar Padilla (Chapter 3, Chapter 7, and Conclusion) were all interviewed in Chile in November 2018 by Matt Kennard. Sergio Cubillos was also interviewed remotely in September 2021 by Tom Gatehouse.

Marino D'Angelo Junior, Maria D'Angelo, and Letícia Oliveira (all Chapter 5) were all interviewed in Brazil in November 2018 by Becky Branford.

Maristella Svampa (Chapter 2); Lucio Cuenca (Chapter 3); Mônica dos Santos, Thiago Alves da Silva, Letícia Aleixo, Marcela Rodrigues, and Vagner Diniz (all Chapter 5) were all interviewed in London on various occasions from 2017 to 2019 by Tom Gatehouse. Thanks to London Mining Network for facilitating where applicable.

All other original interviews were conducted remotely throughout 2021 and early 2022 by Tom Gatehouse and Jo Griffin. We have also borrowed some quotations from online sources where necessary; the original authors are duly accredited.

List of photos

CHAPTER 1
Introduction: No means no

This introduction looks at the origins of the recent community resistance to mining in Latin America, focusing on two cases. It traces back an upsurge of mining conflict in late 2021 in Chubut, in Argentine Patagonia, to a famous case of resistance in the same province from the early 2000s, alongside a case from the north of Peru which has been similarly influential. The chapter also highlights some of the principal means with which communities have opposed the industry, including research, movement building, the call to defend water, and a local direct democracy mechanism known as the consulta popular.

The sound of drums echoes round the sun-drenched streets. There is chanting and whistling; people blow horns. People of all ages stream through the city, from grandparents to young children. A young man pushes a bicycle. Another carries a child on his shoulders. Someone stands wrapped in the Wiphala, the rainbow-coloured banner which represents the indigenous people of the Andes. Argentina's sky-blue and white flag dances above the crowd. People carry placards: 'Chubut won't surrender', 'Mining is plunder: No means no'. Men and women embrace. With the sky darkening, two MCs rap over a simple beat. Later, a cumbia band plays to a crowd which sings their lyrics back at them. Little girls dance in the street.

The day was 21 December 2021; the place, Puerto Madryn, in the province of Chubut, in Argentine Patagonia. The crowds had turned out to celebrate the withdrawal of a 'zoning' law, which would have permitted large-scale open-pit mining in the *meseta central*, an arid, sparsely populated plateau that extends across much of this vast province.

Governor Mariano Arcioni – who had opposed mining in Chubut before coming to power – had got the bill through the legislature in a special session on 15 December, unexpectedly brought forward by a day in order to pre-empt the protests that had been organized by social movements, trade unions, and others. Even so, on the day of the vote there were still violent clashes outside the legislature in Rawson, the provincial capital, between around 300 opponents of the bill and the police.

'The vote was supposed to happen on Thursday, so on Wednesday we all went to Rawson for a vigil, to put pressure on the legislators to vote against the bill or at least to abstain,' says Demián Morassi, an art teacher and member of the Asamblea en Defensa del Territorio Madryn ('Assembly in Defence of the Territory of Puerto Madryn').[1] 'There was going to be music, art … we had planned to spend all night there. But by the time we arrived they were already voting on it.'

'That's when the protests became a bit more violent and aggressive ... The protestors built a roadblock to stop the legislators getting away in their cars, to demand answers and make them explain why they had voted for it. But then the police started firing [rubber bullets] and using tear gas to disperse the crowd and let the legislators out. Some people were injured; others got arrested and taken into custody.'

The violence continued until the early hours of the morning, with the police pursuing protestors through the streets of Rawson. Online videos show people running for cover, as convoys of armed police take pot-shots from the back of pick-up trucks. According to the assemblies, 15 people were arrested and at least 40 injured by rubber bullets and police beatings (Lag and Guerrero, 2021).

'We know that the only way to get mining approved in Chubut is through repression. There is no other option,' said Pablo Lada, a local environmental activist, in an interview on 17 December. 'There was a lot of anger yesterday about what happened the night before last. The repression lasted for hours. In 20 years of activism, I had never seen anything like it. Until 2 a.m. we could still see police on motorbikes, in trucks, dressed in plain clothes, shooting at anyone they found in the street' (La Izquierda Diario, 2021).

On 16 December protests spread throughout Chubut. In Rawson, masked demonstrators were filmed attacking the House of Government and other public buildings, kicking the doors in, breaking windows, starting fires. It remains unclear who they were; the assemblies insist that the resistance to mining in Chubut is peaceful and argue that the vandalism was committed by police infiltrators, as a means of discrediting the movement and justifying further repression (Pandolfi, 2021).

Whoever was responsible, to focus only on the violence in Rawson would be to misrepresent an authentic popular movement with broad civil society support throughout the province: from the scientists at the National Scientific and Technical Research Council (CONICET), to local retailers who announced that the 14 legislators who had voted for the bill would not be welcome in their shops; from dockers who went on strike at the port in Rawson, to all eight Catholic bishops in Argentine Patagonia. Most of all, the movement consisted of ordinary citizens, who took to the streets for five days in a row, marching twice a day in some cases.

On 20 December, Arcioni finally gave in to popular pressure, revoking the new law and promising a referendum on the issue in the future. It had been a successful show of defiance not only towards the provincial government, but also towards the federal government, which had supported Arcioni's attempt to change the law; and towards Argentina's major news outlets, which had largely ignored what was happening and only began reporting on it once the protests turned violent.

'This cause is no longer something that concerns just environmentalists, or the assembly movement,' said Lada. 'What we're seeing is an uprising of the whole community' (La Izquierda Diario, 2021).

Photo 1.1 Demonstrators in Rawson celebrate revocation of the law / © Alex Dukal 2021

Birth of the resistance

The roots of the recent conflict in Chubut go back to 2002, when the American mining company Meridian Gold bought the rights to a gold and silver concession near the town of Esquel, in the foothills of the Andes. El Desquite, the proposed operation, was to produce approximately $1.4 billion dollars' worth of gold and silver over ten years (*Clarín*, 2003).

'Esquel is where the first movement against large-scale mining [in Argentina] emerged,' says Demián Morassi. 'It's in the Andean *cordillera*, a green place where there are national parks, nature reserves, and ski slopes. It's a tourist destination; people go there for the mountains, trees, streams, and waterfalls.'

From an industry perspective, Esquel was always a risky choice of location. It's a place of great natural beauty, to where many people have relocated from larger towns and cities to be closer to nature and enjoy a more peaceful way of life. Local civil society is cohesive, and the population relatively well educated. Moreover, tourism would surely have been impacted by a large open-pit mine located less than seven kilometres outside town (Sohn et al., 2007, p. 27).

Even so, the failure of Meridian Gold in Esquel was not a *fait accompli*. Unlike other countries in the region, particularly its western neighbour Chile, Argentina had little history of industrial mining prior to the 1990s. In 2002, the residents of Esquel – and other towns in Argentina where mining companies were starting to move in – were unfamiliar with industrial mining and its impacts (ibid.).

As a result, the residents of Esquel attempted to get to grips with the issue from very early on. They studied the impacts of open-pit mining and held public meetings at which their findings were presented and discussed. This informal research was backed up by a group of professors at the local branch of the Universidad Nacional de la Patagonia San Juan Bosco, who gave talks in schools and at neighbourhood meetings, promoting the circulation of knowledge on the topic (Marín, 2009, p. 184). The more information locals generated, the more they became convinced of the need to resist.

Meridian Gold also contributed to its own downfall. It did not attempt seriously to engage with the community and failed to address concerns about the project, particularly on the question of the use of cyanide in mineral processing. This created the impression that the company was concealing information and could not be trusted (Sohn et al., 2007, p. 28).

Indeed, the American hydrogeologist Robert Moran, who visited Esquel at the height of the conflict in February 2003, blasted the Environmental Impact Assessment (EIA) Meridian had submitted to authorities as 'the most "undefined" EIA I have reviewed in more than 30 years of hydrogeologic experience.' 'It certainly would not be acceptable to regulators for comparable purposes in western European countries, the U.S.A., or in Canada,' he wrote. 'Much of what is presented is filled with half-truths regarding the details of the processes and potential impacts' (Moran, 2003, p. 3).

Suspicion amongst the local population quickly coalesced into organized resistance and the Asamblea de Vecinos Autoconvocados por el No a la Mina ('Assembly of Self-Convened Neighbours for No to the Mine') was born in November 2002. The first big march against the mine was held on 24 November, followed by another, even larger demonstration on 4 December. In February 2003, recognizing the strength of feeling in the town, local authorities agreed to hold a *consulta popular* – a local referendum – the following month.

On 23 March, the people of Esquel went to the polls. Despite the strong opposition to the mine, local authorities were still hopeful of a 'Yes' vote. Memories of Argentina's economic collapse in 2001 were still raw; unemployment was high, particularly in poorer neighbourhoods; and promises of new jobs and fresh investment in the town must have been appealing for some. Yet the result wasn't even close: 81 per cent voted against the mine. Meridian had little choice but to suspend the project.

Ceviche without lime?

Similar conflicts had been brewing elsewhere in Latin America around the turn of the millennium. Less than a year before the vote in Esquel, locals in Tambogrande, in the north of Peru, had voted on whether to permit exploitation of a large gold, silver, copper, and zinc deposit, by the Canadian company Manhattan Minerals.

Though there are obvious parallels between the two cases – not least the timeframe and the use of the *consulta popular* mechanism – the stakes in Tambogrande were even higher. Manhattan's proposed site for the mine was located right on the edge of the town, and so to make way for the open pit the company planned to relocate 8,000 residents, around half the local population.

It was also a tougher social and political environment for opponents of the project. Manhattan had arrived in Peru in 1993, just as Peru's 1990s mining boom was really beginning to take off – and enjoyed the full support of the national government. Meanwhile, in Tambogrande, most locals had no access to running water or electricity and many were living on less than $2 a day (Hennessy, 2003).

But despite company promises of jobs and new housing and infra-structure, the project generated fierce opposition from the start – and not only amongst residents angry about the proposed relocation. The San Lorenzo valley, in which Tambogrande is located, accounts for much of the national production of mangos and citrus fruit, as well as being home to plantations of rice, grapes, avocados, and other crops. Local farmers were concerned about the impact of mining pollution on their produce, as well as Manhattan's plans to divert a river, in an arid area of Peru where the agriculture depends on careful irrigation (Read, 2017, p. 26).

Another key difference with the Esquel case was the level of violence. In Tambogrande, Manhattan's offices and mining camp were attacked and set on fire, and confrontations took place between locals and police. This came to a head when the farmer, school director, and anti-mining leader Godofredo García Baca was shot dead in March 2001. García's killer was caught and eventually sentenced to 30 years in prison, though he has never admitted responsibility for the crime or revealed if he was acting on anyone's orders. Though there is evidence to suggest it was a targeted assassination, the judges concluded that it was an unpremeditated killing which occurred during a robbery (Thieroldt Llanos, 2017, p. 283).

The murder marked a turning point in the conflict. New actors intervened, including the Catholic Church, national and international NGOs, and the Public Ombudsman. The local resistance adapted its strategy, building a movement based on peaceful protest and seeking alliances with partners elsewhere in Peru and beyond. This eventually led to the *consulta popular*, which was supported by the local government, but only later, reluctantly accepted by the national government and Manhattan Minerals as a means of ending the dispute (Read, 2017, p. 30).

In an appeal to locals' sense of national pride, the 'No' campaign highlighted that the limes grown in the region were an essential ingredient of iconic Peruvian cuisine, such as ceviche and pisco sour. The question 'Can you imagine ceviche without lime?' became a slogan of the movement, and the image of the lime began to appear on flyers, posters, and costumes, sometimes in combination with the Peruvian national colours (ibid., pp. 31–32).

The vote took place on 2 June 2002, with the presence of national and international observers. Nearly 94 per cent voted against the mine.

Photo 1.2 A demonstrator in Tambogrande / Guarango Cine y Video

Despite this unequivocal signal that it was not welcome, Manhattan didn't leave Tambogrande until February 2005, after failing to meet conditions required by the Peruvian state for development of the mine. Nonetheless, the vote had been decisive: a large part of the reason Manhattan failed to honour its contract was that it couldn't find a major partner willing to take on such an unpopular project (Mines and Communities, 2005).

'The issue is water'

Just weeks after the March 2003 *consulta* in Esquel, the government of Chubut passed Law 5,001, prohibiting open-pit metal mining in the entire province, along with the use of cyanide in mineral processing. However, the law did also order the 'zoning' of the province in order to establish where it might be possible to mine and under what conditions this should occur. This was supposed to take place within 120 days of the law coming into force.

It never happened, largely because of popular opposition. But the fact that zoning had been written into the text of the ban left open the possibility that the government might, at a later date, attempt to revive it.

'The law doesn't permit zoning, but the idea has been present ever since it was passed,' explains Demián Morassi. 'That's why they keep coming back to it, saying that yes, large-scale mining is prohibited, but that if it were to happen somewhere where it didn't affect the river basin then it might be possible.'

Photo 1.3 Entrance to the hamlet of El Escorial, Gastre department. The sign reads 'No to large-scale mining. Water is worth more than gold' / © Alex Dukal 2021

For the industry and the provincial government, this 'somewhere' is the departments of Telsen and Gastre on the *meseta central* – the location of Navidad, one of the largest untapped silver deposits on the planet. In spite of Chubut's mining ban, the Canadian mining giant Pan American Silver bought the rights to Navidad in 2010, hoping for a favourable change in the law. The company has been present in the area ever since, carrying out Corporate Social Responsibility (CSR) work with communities and lobbying politicians at both provincial and national level (MiningWatch Canada, 2019; No a la Mina, 2019).

'It was a fabulous deposit in the middle of a perfectly good place to mine, in the middle of a kind of windswept no-man's land in southern Argentina,' said company founder Ross Beaty in a 2020 interview, lamenting the failure to advance the project. 'And it had every technical reason in the world to be a really great place to build a mine. There was no biodiversity loss, no environmental issues at all, no people, no nothing' (Crux Investor, 2020).

But contrary to Beaty's dismissal of the *meseta* as a wasteland, it is home to an agricultural and pastoral population, including indigenous Mapuche-Tehuelche communities. These two groups native to southern Argentina and Chile have had a long and painful history of conflict, first with the Spanish, then with the two independent states that followed colonial rule. They have taken a stand against the zoning proposal and are demanding that their right to consultation be upheld, as required by Convention 169 of the International Labour Organization (ILO), to which Argentina is a signatory (see Chapter 3).

'In general, these traditional communities are anti-mining, because they depend on tiny streams of water. It's true that the aquifer is large, but these streams are small,' says Morassi. 'It's where they take their sheep to drink, it's where the guanacos drink, on which they depend for food, as well as all the other little animals they eat – hares, rodents – and the crops they plant, everything depends on these little streams.'

'So they see mining as a threat to their entire history. There are Mapuche-Tehuelche communities there that are hundreds of years old, and they're the ones who most defend these little streams and springs.'

But it was not only Chubut's indigenous communities who were concerned about the potential impacts of Navidad on their water supply.

'This is an enormous province, which has uranium, gold, silver ... a whole load of minerals. But we only have one major river [the Río Chubut], which supplies the entire province with water,' explains Morassi. 'And this is a very dry region. We depend on the river for everything. Sometimes there's no water. If there isn't much snowfall in the winter, the river runs low, and there are shortages.'

Precipitation has been below average in recent years, leading the provincial legislature to declare a 'water emergency' in August 2021. The very same day the zoning bill was approved, there were water shortages in Comodoro Rivadavia, Chubut's most populous city. Experts have warned that climate change is likely to put further pressure on water supply in the years to come; they predict a 40 per cent reduction in the flow of the Río Chubut by the end of the century (CONICET-CENPAT and Lab EcoFluvial, 2020).

In this context, it's no surprise that the decision to permit large-scale mining – an industry which can consume tens of millions of litres of water a day – had people up in arms across the province, from the Andean *cordillera* to the Atlantic coast.

'The defence of water is what has blocked mining in Chubut,' says Morassi. 'The issue is water, that's the framework we're operating in.'

The Esquel effect

Both Tambogrande and Esquel have been an inspiration to other communities across Latin America resisting the mining industry. Over the last two decades or so, communities all over the region have adopted many of the same tactics.

These include horizontal forms of organization, as the best way to ensure support from across the community. There is often an attempt to dispute specialist knowledge, involving independent research and the production and sharing of information, sometimes with support from academics and NGOs. Mining opponents also tend to emphasize the need to preserve local livelihoods and traditions, particularly in agricultural communities. Finally, there is almost always a call to defend the water supply.

One strategy which has been deployed elsewhere to great effect is the *consulta popular* (and not only in relation to mining; referenda have also been held on hydrocarbon extraction, as well as large infrastructure projects). There have been dozens of *consultas* in towns across several Latin American countries, including Guatemala and Ecuador, as well as elsewhere in Peru and Argentina.

In Colombia alone, there have been a total of eight *consultas* on mining, six of which were held in 2017. In all eight cases, at least 96 per cent voted no to the mine (González, 2019). By early 2018, there were another 54 *consultas* in the pipeline, mostly on mining and hydrocarbons, a situation which had investors rattled and was threatening to cause a major headache for the Colombian state. Ultimately, the Constitutional Court intervened, ruling in 2018 that the national state was the owner of subsoil resources; and then in 2019 that there was no legal requirement for a *consulta popular* on projects which imply a significant change in the use of land.

Nonetheless, while the legal force of the *consultas* may vary, a thumping 'No' vote is sometimes enough to deter mining companies and their investors.

'I think the true power of the *consultas populares*, beyond any legal effects, lies in their social and political effects,' says David Fajardo Torres, an Ecuadorian law student and environmental activist who campaigned on a *consulta popular* which blocked mining near the city of Cuenca (see Chapter 6). 'To organize and then win a *consulta popular* gives your resistance movement an indisputable legitimacy. Today, absolutely nobody can say that Cuenca is in favour of mining, because the results speak for themselves.'

In Argentina, the movement in Esquel sparked off a nationwide wave of community resistance to mining, sometimes called the 'Esquel effect', which led eight other provinces to follow Chubut's lead and introduce tight restrictions on the industry (though in two cases the laws have since been revoked). Today, mining in Argentina is mostly concentrated in just two provinces: Santa Cruz, Chubut's southern neighbour, and San Juan (see Chapter 3), though there are rich mineral deposits all over the country. This 'Esquel effect' has been key.

One example comes from the province of Mendoza, further to the north. 'Mendoza has some lucrative mineral deposits which haven't been explored due to the opposition,' says Lucrecia Wagner, a researcher at the National Scientific and Research Council (CONICET) in the provincial capital. 'Besides which, the mining sector freely admits it's an attractive province in terms of the services it can provide. The city of Mendoza is the fourth biggest in the country and it has the best transport links to Chile. In terms of logistics, it would be of huge benefit to the mining sector if it were able to operate here.'

But since 2007, Mendoza has prohibited the use of toxic substances in mining processes, effectively making industrial metal mining impossible in the province. Local civil society is quick to mobilize in defence of the ban whenever it comes under threat, most recently in late 2019, when the government passed another law overriding it. With obvious parallels with what was to happen two years later in Chubut, ten days of constant street demonstrations forced the government to strike down the new law.

At the national level, the Central American countries have been leading the way. In 2010, Costa Rican lawmakers voted unanimously to ban open-pit mining and the use of cyanide and mercury in mineral processing. Mining opponents made the case that the industry would have threatened Costa Rica's spectacular biodiversity, trashing the country's reputation as a global leader in environmental conservation. Ninety per cent of the population supported the ban (Salva la Selva, 2010), with mining widely seen as incompatible with a model of development based historically on agriculture and increasingly on tourism, now Costa Rica's main source of income and hard currency (Embajada de Costa Rica en Washington DC, n.d.).

In March 2022, a new, progressive administration in Honduras introduced a similar ban, in a major victory for frontline communities and environmental activists, after more than a decade of often harrowing struggle. Following a military coup in 2009, Honduras had been aggressively carved up for the benefit of large-scale agribusiness, major tourist developments, *maquila* manufacturing, hydroelectricity generation, and mining. During this time the country became one of the most dangerous places in the world for environmental and human rights defenders; many of those who resisted were harassed, threatened, beaten, imprisoned, and in some cases, killed (see Chapter 4).

But no other country has yet gone as far as El Salvador, the smallest mainland country in the Americas, where, after a nationwide campaign by communities, activists, NGOs, the Catholic Church, and others, in 2017 legislators passed a blanket ban on all metal mining – both open-pit and underground – and its related activities (see Chapter 2). El Salvador is suffering from acute water stress, and again, the need to defend its remaining freshwater resources was crucial in building momentum for the ban.

'This is a historic, brave, and momentous decision,' said Guillermo Gallegos, then president of El Salvador's legislative assembly. 'The territory of our country is very small and to allow metal mining would practically be suicide' (Paullier, 2017)

***

The events of December 2021 in Chubut are a reminder that mining conflicts are rarely simple cut-and-dried affairs. Victories for communities and social movements may not be definitive; sometimes they have to be defended. Conflicts are shaped to a significant extent by market fluctuations, with the pressure from the industry becoming stronger at moments of high mineral prices. Time also tends to favour the companies, which, like Pan American Silver in the case of Navidad, often play the long game, trying to engineer a change of circumstances which would permit their operations to advance.

Even in Esquel, the industry has not given up hope. In 2007, Meridian Gold was taken over by the Canadian giant Yamana Gold, which renamed the El Desquite project 'Suyai' and repackaged it as a small-scale underground mine, with a design 'intended to be respectful of and deal with local community concerns' (Yamana Gold, 2022). As recently as 2020, Yamana granted an Argentine asset management company the right to acquire a 40 per cent stake in the project, in return for taking care of environmental, social, and governance (ESG) issues – in other words, for engaging with local authorities and civil society to try and advance the project. Given Chubut's mining ban and the strength of local opposition, this is hard to envisage – but it remains a possibility.

We should also not romanticize resistance to mining; it can be a messy, complicated, and bad-tempered business. It is not easy to defy companies of such immense power and wealth, nor governments which are bent on facilitating their operations. Resistance may involve great personal sacrifice: not simply time and energy, but also jobs, homes, relationships, and even, in some cases, lives.

The enormous scale of open-pit mining operations also tends to galvanize resistance from a wide range of social sectors, which can make for some fractious and uneasy alliances between groups of diverse concerns and motivations. Resistance may encompass environmentalist or anti-capitalist activists, and capitalist farmers who produce crops for export; small business-people disgruntled that a mining company has not contracted their services, and local politicians who piggyback on popular resistance as a means of furthering their careers (Bebbington et al., 2013, p. 279). Yet conversely, it is often this very diversity which endows successful resistance movements with their strength and legitimacy.

'Though we don't compromise on our principles and political ideals, we mustn't be purists, in the sense of refusing to deal with people who think differently,' says David Fajardo Torres. 'That would get us nowhere. At the end of the day, it's through dialogue with people who don't think like us that we can bring about these processes of change.'

In many cases, resistance is simply a question of self-preservation: a struggle to guarantee ongoing access to drinkable water and a healthy environment, and beyond that, to ensure that one's community will continue to exist in the same place and remain habitable for future generations. Alongside this, in recent years, there has increasingly been an attempt to articulate alternative visions, both for communities and wider societies, based on the conservation and sustainable management of land, water, shared resources, and nature in general (Svampa, 2019, p. 12).

Whatever the motivation, resistance to mining is not something to be undertaken lightly. It is often slow, laborious, and, depending on the context, may involve considerable personal danger. But where it has been successful, it provides remarkable examples of what communities and social movements can do when they unite around a common cause.

Note

1. In Argentina, the assemblies are neighbourhood organizations which emerged during the country's economic collapse in 2001. At first, these were concerned largely with local issues; over time, many assemblies have come to focus on broader socioenvironmental concerns and the threats posed by major enterprises, not only mining, but also agribusiness, road building, manufacturing, tourist developments, and others. They are characterized by non-partisanship, horizontality, and principles of direct democracy.

References

All references to web-based material were checked and still available in November 2022 unless otherwise stated.

All references are listed, with clickable links for your convenience, on the page for this chapter on the Heart of Our Earth website: <https://lab.org.uk/the-heart-of-our-earth/>

Bebbington, A., Bury, J. and Gallagher, E. (2013) 'Conclusions'. In: A. Bebbington and J. Bury, eds., *Subterranean Struggles. New dynamics of mining, oil, and gas in Latin America*. Austin, TX: University of Texas Press, pp. 267–288.

Clarín (2003) 'Rechazo a un proyecto minero en Esquel.' [online] Available at: <https://www.clarin.com/opinion/rechazo-proyecto-minero-esquel_0_SkegAxMeCtx.html?code=bxb6li_4HHVFsLmVs5YKu0LUeziB69a6LLIIScuB41Egs&state=RVdvUTF2aGVkdGIxdHJ0VW44RXRJR05pM2hNUzZNLVhybHlwa0tNUmEzQQ==>.

CONICET-CENPAT and Lab EcoFluvial (2020) *Un río, todas las aguas.* [online] Available at: <https://unriotodaslasaguas.com.ar/wp-content/uploads/2021/12/Resumen_esencial_2021.pdf>.

Crux Investor (2020) *So You Want to be an Entrepreneur?*. [video] Available at: <https://www.youtube.com/watch?v=SNYgcBOlJWQ>.

Embajada de Costa Rica en Washington DC (n.d.) 'About Costa Rica'. [online] Available at: <http://www.costarica-embassy.org/index.php?q=node/19>.

González, X. (2019) 'Comunidades votaron en 10 consultas populares mineras desde el 2013'. [online] *La República*. Available at: <https://www.larepublica.co/especiales/minas-y-energia-marzo-2019/comunidades-votaron-en-10-consultas-populares-mineras-desde-el-2013-2842036>.

Hennessy, H. (2003) 'Un pueblo peruano no cambia frutas por oro'. [online] BBC Mundo. Available at: <http://news.bbc.co.uk/hi/spanish/business/newsid_3289000/3289583.stm>.

Lag, N. and Guerrero, M. (2021) 'Zonificación minera: trampas legislativas y represión policial para aprobar la megaminería en Chubut'. [online] Tierra Viva. Available at: <https://agenciatierraviva.com.ar/zonificacion-minera-trampas-legislativas-y-represion-policial-para-aprobar-la-megamineria-en-chubut/>.

La Izquierda Diario (2021) *Chubut: protestas y represión: Entrevista con Pablo Lada, activista ambiental*. [video] Available at: <https://www.youtube.com/watch?v=ab84h8cj6EQ>.

Marín, M. (2009) 'El "no a la mina" de Esquel como acontecimiento: otro mundo posible'. In: M. Svampa and M. Antonelli, eds., *Minería transnacional, narrativas del desarrollo y resistencias sociales*, 1st ed. [online] Buenos Aires: Biblos, pp. 181–204. Available at: <http://maristellasvampa.net/wp-content/uploads/2019/12/Miner%C3%ADa-transnacional.pdf>.

Mines and Communities (2005) 'Manhattan pulls out after US$60mn Tambogrande loss – Peru'. [online] Available at: <http://www.minesandcommunities.org/article.php?a=7530>.

MiningWatch Canada (2019) 'Canadian Mining Lobby Pushes to Amend Environmental Legislation in Chubut Province, Argentina'. [online] Available at: <https://miningwatch.ca/news/2019/11/26/canadian-mining-lobby-pushes-amend-environmental-legislation-chubut-province>.

Moran, R. (2003) *Esquel, Argentina. Predictions and promises of a flawed Environmental Impact Assessment.* [online] Greenpeace Argentina/Mineral Policy Center. Available at: <https://earthworks.org/assets/uploads/archive/files/publications/PredictionsPromisesFINAL.pdf>.

No a la Mina (2019) 'Pan American Silver y su careta de benefactor'. [online] Available at: <https://www.ocmal.org/pan-american-silver-y-su-careta-de-benefactor/>.

Pandolfi, F. (2021) 'El Chubutazo y la sociedad en movimiento: ¿Cómo se ganó?'. [online] lavaca. Available at: <https://lavaca.org/notas/chubutazo-y-la-sociedad-en-movimiento-como-se-gano/>.

Paullier, J. (2017) '"Un día histórico": cómo El Salvador logró prohibir por ley la minería metálica en el país'. [online] BBC Mundo. Available at: <https://www.bbc.com/mundo/noticias-america-latina-39451498>.

Read, A.B. (2017) 'Defending Home: How Resistance Movements are Framed Against Mineral Extraction in Cajamarca and Tambogrande, Peru'. In: *International Development, Community and Environment (IDCE)*, [online] 105. Available at: <https://commons.clarku.edu/cgi/viewcontent.cgi?article=1173&context=idce_masters_papers>.

Salva la Selva (2010) 'Costa Rica prohíbe la minería de oro a cielo abierto'. [online] Available at: <https://www.salvalaselva.org/exitos-y-noticias/3196/costa-rica-prohibe-la-mineria-de-oro-a-cielo-abierto>.

Sohn, J., Herz, S. and LaViña, A. (2007) *Development without conflict: The business case for community consent.* [online] World Resources Institute. Available at: <https://sarpn.org/documents/d0002569/index.php>.

Svampa, M. (2019) *Las fronteras del neoextractivismo en América Latina. Conflictos socioambientales, giro ecoterritorial y nuevas dependencias.* [online] CALAS. Available at: <http://www.calas.lat/sites/default/files/svampa_neoextractivismo.pdf>.

Thieroldt Llanos, J. (2017) *The Local Dimension of Transnational Activity in Environmental Conflicts: Tambogrande, 1961–2004.* Ph.D. University of Kansas. Available at: <https://kuscholarworks.ku.edu/bitstream/handle/1808/26472/Thieroldt_ku_0099D_15434_DATA_1.pdf?sequence=1&isAllowed=y>.

Yamana Gold (2022) 'Portfolio - Strategic Assets'. [online] Available at: <https://www.yamana.com/English/portfolio/strategic-assets/default.aspx>.

CHAPTER 2
From old to new mining

This chapter provides an overview of two historical mining cycles in Bolivia: silver, focusing on the Cerro Rico in Potosí; and tin, focusing on Simón Patiño and the city of Oruro. It then analyzes mining in Latin America since the 1990s, showing how this has taken on shapes and dimensions fundamentally distinct from those of mining in earlier periods. It then scrutinizes the role of the Canadian mining industry and the Canadian state in Latin America in recent decades. Finally, it tells the story of a famous investor-state-dispute-settlement (ISDS) lawsuit, brought by the Canadian company Pacific Rim against El Salvador.

I. The original sin

In April 2020, with Covid-19 cases on the rise in Bolivia and a full nationwide lockdown in force, around 10,000 miners in the city of Potosí reluctantly downed their tools.

The pandemic had achieved something that not war, revolution or even other outbreaks of disease had ever managed: it brought a complete halt to mining activity on the Cerro Rico – the picturesque conical mountain overlooking the city, which appears on the Bolivian coat of arms. This is the only time on record that this has happened since the Spanish Empire began silver production in 1545 (Harris, 2020).

By far the richest source of silver in the history of mankind, the Cerro Rico is often cited as a symbol of mining in Latin America, most famously by the Uruguayan writer Eduardo Galeano in his influential text *The Open Veins of Latin America*. It vividly evokes the region's deal with the devil: on the one hand, fabulous wealth, power, and opulence; on the other, decline and destitution, disease and death (Brown, 2012, p. 197).

Founded as a remote mining settlement of just 3,000 inhabitants in the 1540s, Potosí had become a city of as many as 160,000 by the early 17th century – a population larger than London, Milan, or Seville at the time. The Cerro Rico made the Spaniards who controlled production immensely wealthy and bankrolled the expansion of the Spanish Empire, providing the funds for wars against the British, Dutch, French, and Ottomans. At the same time, countless thousands of indigenous *mitayos* (Indians forced to provide labour as tribute to the Spanish) and African slaves were sacrificed to Europe's voracious appetite for silver (Greenfield, 2016).

Yet by the end of the 17th century the best ore had been exhausted and the city's population had fallen to just 60,000. Today, Potosí is the poorest department in what had long been, until recently, South America's poorest country.[1] According to official statistics, as of 2018, almost a third of the department's nearly 900,000 inhabitants were living in extreme poverty (Instituto Nacional de Estadística, 2022).

But more than four centuries since its heyday, miners continue to work the Cerro Rico, using drills and dynamite to extract whatever is left of value in its depths: not only silver, but also lead, zinc, and tin. They belong to associations known as cooperatives, which emerged in the 1980s following privatization and mass layoffs at Bolivia's state-owned mines. The term is a misnomer: the bosses take the lion's share of the profits, while miners are responsible for their own production, labouring in conditions which remain precarious in the extreme. They have no health insurance, pensions or other benefits, and health and safety regulation is virtually non-existent. Most die young, if not from an accident or gas poisoning, then from silicosis, an incurable lung disease resulting from prolonged exposure to silica dust. The average life expectancy for a Cerro Rico miner is just 40.

There is also industrial private sector mining on the Cerro Rico by Empresa Minera Manquiri (EMM), currently a subsidiary of the Canadian company Andean Precious Metals. Since 2008, it has mined surface deposits on the Cerro Rico, as well as purchasing silver oxides from cooperatives working underground (MiningWatch Canada, 2021).

Ravaged by almost 500 years of constant mining activity, the Cerro Rico is now in danger of collapsing altogether. In January 2011 a sinkhole appeared at the summit. Engineers from state mining company COMIBOL patched up the damage with ultra-light cement, but the risk of collapse remains. Videos online show columns of dust rising up from the summit, a sign that the mountain continues slowly to implode. As of late 2021, there are twelve sinkholes at various points on the mountain. It is now illegal to operate above the 4,400-metre mark, but the miners – both the cooperatives and, according to the Potosí Civic Committee, EMM – flout the law (Francescone, 2021).

For Kirsten Francescone, former Latin America Program Coordinator at the NGO MiningWatch Canada, it is the presence of EMM, more than the cooperatives, which has brought the Cerro Rico to the verge of collapse. Its use of intensive methods on the Cerro Rico's superficial deposits has damaged the mountain's structure, while its purchase of mineral from cooperatives working in unstable areas – such as above 4,400 metres – has encouraged more of them to do so, increasing their production to unprecedented levels (ibid.).

Despite the threat of a major disaster, the miners say they have no choice. Potosí has always been a mining town; there are few opportunities outside the sector. Many of the family members of the miners who work the Cerro Rico also work in mining in some way. Boys as young as ten go into the mines with their fathers or other family members to help out when not at school. Women participate too: aside from their work as homemakers, they help to process minerals – washing and drying them, for instance. Some work as *palliris*, manually sifting through the spoil heaps outside the mines for any ore of value, while single women and widows guard mine entrances – a cold, lonely, and dangerous job, given that they live high up on the hillside, often with just a dog for company.

Potosí represents Latin America's original sin. Its silver drained away to Spain, to northern Europe and beyond, while little of the capital obtained was reinvested in other activities that might have been more dynamic and sustainable, or which might have provided safer employment and better quality of life (Brown, 2012, p. 43). Along with silver from Mexico and sugar from Brazil, it established Latin America as a provider of raw materials to global markets – a relationship which persists to this day.

Unsurprisingly, the shutdown at the Cerro Rico due to the pandemic lasted little more than a month. With the miners risking their lives every day to scrape out a living from within the depths of the mountain, what fear could Covid-19 possibly hold for them?

From tin to rust

On the outskirts of the city of Oruro – 234 km northwest of Potosí, across the *altiplano* – there is a curious piece of public art.

It is a giant miner's helmet of tin and burnished metal, mounted on columns in the centre of a large traffic roundabout. Where the headlamp would normally rest, there is a window, open onto an image of La Virgen del Socavón, or 'Our Lady of the Mineshaft'. The patron saint of Oruro, this image of the Virgin Mary is venerated by miners, who pray for her protection while they labour underground.

Arranged around the helmet there are sculptures of a snake, a lizard, a toad, and ants, representing four plagues sent to the local Uru people by the god Wari, to punish them for their worship of Pachacamaj, represented by Inti, the sun god. But the Urus are saved by a mysterious Ñusta – an Incan princess – all dressed in white, who turns the plagues to stone or sand. Today, the Ñusta is equated with La Virgen del Socavón.

The artwork is a fine example of the fusion of indigenous and Catholic imagery often found in Andean culture. But above all, it is a tribute to the history of a city which was at the centre of Bolivia's second great mining boom: namely, tin, which was the main national export for most of the 20th century. By 1900, tin had surpassed silver, accounting for more than half of Bolivia's export revenue (Britannica, n.d.).

That same year, a hitherto unknown mining entrepreneur called Simón Patiño discovered a particularly rich vein of tin called La Salvadora, in Llallagua, about 65 km from Oruro. This was the first piece of an empire which was to consist of around half of Bolivia's tin-producing operations, smelters in England and Germany, and further mines in Malaysia. By 1925, Patiño – a mestizo from Cochabamba who had grown up in poverty – was a multimillionaire with an annual income greater than the Bolivian national budget. When he died in 1947 he was one of the richest men in the world.

Photo 2.1 The miner's helmet of Oruro / Juan revolver 2013 / CC BY-SA 3.0

'When tin mining was at its height, Oruro was the most developed city in Bolivia,' says Héctor Córdova, a mining expert and former COMIBOL president. 'It was the first city to have a sewerage system, an electricity network, telephone lines … It was declared the industrial capital of Bolivia, because until around 1930 or 1940 there were three thousand factories operational, as well as smelters and various mines located nearby.'

Tin extraction required much higher levels of capital investment than silver, and so Patiño and Bolivia's other 'tin barons' quickly sought alliances with capital from Europe and the United States. And with migration from Europe increasing in the wake of the First World War, Oruro soon became a dynamic and cosmopolitan city.

'There emerged a certain bourgeoisie of European origin who were very strong economically, linked to the mining sector. But they developed industries related to their origins,' says Córdova. 'For example, the Italians made pasta. The Yugoslavians set up shoe factories, they were very skilled at working with leather. The Spanish set up businesses and shops, selling products they imported from Europe and distributing them throughout the country.'

'So Oruro became a nerve centre of the Bolivian economy. Practically half the economic activity of the country revolved around what was happening in Oruro.'

Patiño is a controversial figure in Bolivia. Many consider him little better than a parasite who grew fat on the country's natural resources and labour, while giving back little in return (Brown, 2012, p. 128). Workers at his mines endured abysmal pay and conditions and were brutally repressed when they fought for improvements, such as at the Catavi Massacre in 1942, when an unknown number – likely to be in the hundreds – of striking miners and their family members were killed after soldiers opened fire on a demonstration.

Such violence only served to radicalize the miners further. In 1952 there was a revolution. Workers and campesinos came together to overthrow the oligarchy of large landowners and tin barons, and the miners were key to its success: the Bolivian Mineworkers Union, which had emerged from the mining camps of Oruro, was on the frontline of the battle against the military. One of the first actions of the new government was the creation of COMIBOL and the subsequent nationalization of the mines held by Patiño and the other tin barons. COMIBOL remained the principal source of revenue for the Bolivian state until the 1980s, when falling prices sent the tin industry into decline (Olave, n.d., p. 4).

Today, many of the former tin mining settlements have become virtual ghost towns; much of the old infrastructure has been left to rust (Brown, 2012, p. 194). Mining cooperatives continue to work the old deposits, though with no access to the machinery or capital of the few remaining state-run or privately held mines in Bolivia, they are limited to primitive methods, with little or no environmental control of their activities.

Since the decline of the mines, Oruro has fared better than Potosí; its strategic location has seen it become a hub for haulage companies operating throughout South America. Still, it is a shadow of its former self, a rather grey and desolate place. The city centre is choked with traffic. Women in traditional dress sift through piles of rubbish looking for anything of value, accompanied by dirty, dishevelled children. Stray dogs forage for scraps.

II. The new mining

One might think the history of Bolivia would serve as a cautionary tale. But since the 1990s, mining has flourished throughout Latin America, as have other so-called 'extractive' industries. These are defined as 'economic activities that remove a natural resource from the environment, submit it to marginal or no processing, and then sell it on' – so not just mining and oil and gas extraction, but also timber, as well as monoculture cash crops such as soya, sugarcane, and palm oil (Bebbington, 2010, p. 97).

Prices for many of these commodities increased steadily from the 1990s until around 2014, in a period known as the commodities supercycle. This was fuelled by demand from the rapidly growing Asian economies, particularly China, as well as the expanded consumption of the upper and middle classes around the world (Tetreault, 2015, p. 50). Mineral prices enjoyed a particularly steep rise from 2000; prices tripled between 2003 and 2011 (Ballard et al., 2012, p. 55).

For mining investors, the 1990s and 2000s were a period of extraordinary profitability and unprecedented new opportunities. From 1990 to 1997, foreign direct investment (FDI) in mining exploration – the search for new mineral deposits and the assessment of their viability – increased 90 per cent worldwide. But in Latin America it increased by 400 per cent and in Peru by a staggering 2,000 per cent (Bebbington and Bury, 2013, p. 15).

It's not only traditional 'mining countries' which have participated in this bonanza: FDI has also flowed into countries with little or no prior history of industrial mining, such as Argentina, Ecuador, and the Central American countries. And even in countries of long mining tradition, such as Mexico, Peru, and Chile, investment has often funded exploration in entirely new areas, including remote locations of difficult access that the industry tended to overlook in the past, such as high in the Andean *cordillera* or deep in the Amazon rainforest (ibid., p. 16).

'The advance of mining in Latin America has been very intense, in a very short period of time historically,' says Lucio Cuenca, an engineer and director of the Latin American Observatory of Environmental Conflicts (OLCA), based in Santiago de Chile. 'We're talking about countries which have developed the industry over just two or three decades, countries which had no mining, but which have now been completely invaded by it.'

Furthermore, technological advances have permitted the profitable extraction of much lower-grade ore than was previously viable. Some open-pit goldmines remove a tonne of rock to extract less than a single gram of gold, for example. For this to be profitable, enormous quantities of rock have to be removed and processed, which means the wholesale transformation of landscapes in ways which would have been unthinkable in the days of the precarious underground shafts (Bebbington et al., 2013, pp. 270–271).

Not only does this require the use of vast amounts of water and energy, it also generates mountains of waste, or tailings. Tailings are a mixture of water, rock from which any minerals of value have largely been removed, and residues of any chemicals used in processing the ore (Kuyek, 2019, p. 30). They pollute the surrounding environment unless properly contained, usually behind a dam; these structures require careful long-term monitoring beyond the end of a mine's active life, to ensure they are sound and there is no seepage or danger of collapse. But communities located close to dams often feel a sense of mortal danger, a fear much heightened by two major tailings dams failures in Brazil in recent years (see Chapter 5).

It's no surprise that this new mining has provoked a backlash from affected communities, which feel that their traditional ways of life – sometimes going back many generations – may be destroyed by a single business venture with a lifespan of as little as twenty or thirty years (Machado and Fachin, 2015, p. 63). Usually, people from these communities are asking for no more than the right to decide what happens to the territory where they live and upon which they depend. But articulating this demand has often seen them become the target of repression from state and para-state forces acting in the interests of the mining industry (see Chapter 4).

Legalized plunder

In many respects the early 1990s were a time of great hope for Latin America. Across South America, the dictatorships of the Cold War had fallen, giving way to modern democratic systems of government. In Central America, war was coming to an end. A ceasefire was agreed between the Sandinistas and the Contras in Nicaragua in 1990, and peace accords brought the Salvadoran Civil War to a close in early 1992.

Progressive new constitutions came into force, for example in Brazil (1988) and Colombia (1991), which expanded civil liberties and strengthened state institutions. Significantly, hitherto marginalized groups – such as indigenous people and those of African descent – were granted official recognition, and the right to pursue their own cultural lifestyles without pressure to assimilate.

But the 1980s had been a bruising decade. Lavish borrowing during the 1960s and 1970s – in some cases by unelected military regimes – had led to a crisis which sank much of the region into a deep recession. Unable to service their debts, Latin American countries were forced to accept rescue packages devised by the International Monetary Fund (IMF), central banks, and commercial banks, and led by the United States. They were permitted to restructure their debts and were lent the funds to pay the interest – but this help came at a price.

As instructed by the IMF, World Bank, and U.S. Treasury, Latin American governments began to implement a sweeping programme of market-oriented reforms. Publicly owned companies and other assets were sold off – often at knockdown prices – while public sector employees were made redundant. Spending on health, education and social programmes was slashed, while public works were left unfinished. This set of policy prescriptions, which later became known as the Washington Consensus, took a major toll on Latin American society, with high unemployment, steep falls in income, and stagnant economic growth. Not for nothing have the 1980s become known as Latin America's 'lost decade'.

At the same time, countries were under pressure to depart from protectionism, liberalize trade, and open up their economies to foreign investment. This was based, above all, on the intensive extraction of raw materials and their export to global markets. This process has been called the 'reprimarization' of Latin America's economies, following state-led attempts to industrialize during the post-war period.

In the mining sector, it meant the privatization of state-owned companies and mineral deposits, along with the removal of barriers to foreign capital and requirements for companies to hire local workers and source local goods and services. Generous incentives for investment were created, while environmental protections were reformed. States also built new infrastructure to facilitate extraction and provided military and police support to guarantee security at mine sites (Bury and Bebbington, 2013, pp. 44–45).

Taxes on mining operations in this period have generally been very low, as have royalty payments – a tax paid as compensation for the loss of a non-renewable resource. Moreover, they are often accompanied by loopholes and legal flexibility, meaning that companies are able to reduce their responsibilities to the bare minimum. Even the Organisation for Economic Cooperation and Development (OECD) has recommended abolishing the preferential tax treatment of mining and hydrocarbon extraction, both on economic and environmental grounds (MiningWatch Canada, 2005).

Tax is particularly important in the case of mining, given that once metals and minerals have been sold and shipped out of the country, they are gone forever – along with any taxes or royalties that the miner has managed to evade. What remains are landscapes which are irreversibly transformed and sometimes severely polluted, unless there are comprehensive measures in place to ensure responsible mine closure and rehabilitate former mining areas, which is rare (Grupo de Trabajo sobre Minería y Derechos Humanos en América Latina, 2014, p. 20).

In Peru, for example, there are nearly 8,500 hazardous mining waste sites throughout the country which the state has been unable to clean up, either because the companies responsible cannot be held legally accountable (in many cases they no longer exist), or because the damage is irreversible. Some of them date back to the colonial period (Servindi, 2020).

Progressive extractivism?

Many theorists who have studied this new mining in Latin America make a distinction between 'extractivism' and 'neo-extractivism': the former epitomized by conservative or neoliberal governments such as those of Álvaro Uribe in Colombia, Vicente Fox and Felipe Calderón in Mexico, and Alberto Fujimori in Peru; the latter, by the governments of the so-called 'Pink Tide', the wave of left-wing or centre-left governments which came to power across the region following the election of Hugo Chávez in Venezuela in 1998.

The term 'Pink Tide' encompasses a range of governments which were far from uniform: from the more radical 'Bolivarian' governments of Venezuela, Ecuador, and Bolivia; to the centre-left administrations of the Brazilian Workers' Party (PT), to the fairly centrist Concertación and Nueva Mayoría administrations in Chile, for example. Nonetheless, they shared a number of common features. They marked a rupture – some much more than others – with the neoliberal economics of the 1980s and 1990s, establishing a more proactive role for the state (Machado and Fachin, 2015, p. 61). They all attempted to address longstanding social problems exacerbated by the lost decade of the 1980s, and to attend to the needs of the most marginalized sectors of society, particularly the urban poor.

Yet there was little attempt to revive traditional ideals of the Latin American left like nationalization of key industries, land reform, and economic independence. In fact, the Pink Tide governments displayed some striking similarities with what had preceded them, and with what was happening in neighbouring countries with conservative governments: deeper insertion into global markets and openness to foreign investment, mostly in the extractive industries (ibid.).

'There had never been so much money sloshing around in Latin America, it was unprecedented,' says the Argentine sociologist Maristella Svampa. 'The temptation to just continue milking this cash cow, rather than diversifying, or taking on the major economic sectors, was intense. No Latin American country managed to carry out fiscal reforms that would have had an impact on the most powerful sectors, nor did they even really try.'

What made neo-extractivism distinct was, firstly, that the Pink Tide governments attempted to strike a more favourable deal with global capital and secure a greater slice of the profits for the state. Secondly, some of this revenue was redistributed more equitably, through social programmes and poverty relief.

Some of the initial results were undeniably impressive. Poverty rates in these countries tumbled, much faster than in Latin American countries with non-Pink Tide governments. Bolivia almost halved its poverty rate in just over a decade (2006–17) (Balch, 2019), while more than 30 million Brazilians are believed to have risen out of poverty during Lula's presidency (2003–11).

Even so, while the more progressive governments have sometimes managed to negotiate a better deal with companies in the extractive industries, the state's share of exported natural resource wealth does not usually exceed ten per cent (Petras and Veltmeyer, 2018. p. 33). The real winners remain those located in the global networks which facilitate extraction, principally the companies and their investors, but also other stakeholders, such as banks, financial institutions, law firms, geological and mineral consultants, and others.

Like the great cycles of silver and tin in Bolivia, the commodities supercycle of the early 21st century came to an end. By 2015, Latin America was in recession. Many of the social advances which had been built on the shifting sands of the boom have turned out not to be sustainable even in the medium term. In the years since, many of those who were once hailed as Latin America's 'new middle class' have fallen back into poverty – a trend exacerbated by the Covid-19 pandemic, which has hit Latin America harder than any other global region.

Despite the end of the boom, extraction has continued apace and has even expanded, in order to cover the trade deficit created by lower export prices. As well as ever more open-pit mining, this includes so-called 'extreme energy' exploration; in other words, the development of energy sources requiring even more intensive and environmentally destructive methods of extraction than conventional fossil fuels, such as shale (fracked) gas, tar sands oil, and pre-salt hydrocarbons.

Now, as Latin American governments struggle to plug the holes that Covid-19 has blown in national budgets, the temptation once again is to expand the extractive frontier further still, to tap the lucrative rents that these industries can provide. Mining will be at the forefront of this, with many analysts predicting a huge spike in metal and mineral prices in the coming years due to the transition away from fossil fuels and towards renewable energy and cleaner technologies, which are metal and mineral intensive (see Chapter 7).

As much as ever, Latin America remains in thrall to its original sin.

III. A secondary imperialist power?

Following independence, elites throughout Latin America identified mining as one of the means with which to build modern, civilized nations – a view which persists today in the emphasis on mining as a path to 'development' (Grieco, 2016, p. 131). Initially, the industry was dominated by British companies and investors, and then, from the late 19th century onwards, the Americans. Today, there are still major Latin American mining assets in the hands of companies listed in the UK and the United States, while both the London and New York stock exchanges remain major hubs for mining investment.

However, the new mining in Latin America has a far more global character than in the past. Operating mines and significant claims are held by companies from all over the world: Australia, Japan, South Africa, India, and increasingly China. The presence of Latin American companies is also significant: Grupo México (see Chapter 5) is a major player, not only in Mexico, but also in Peru, via its subsidiary Southern Copper Corporation. The Brazilian company Vale (see also Chapter 5) has a presence in 27 countries worldwide and dominates the Brazilian mining sector, particularly in iron ore. Finally, the world's leading copper miner by production is the Chilean state-owned company CODELCO.

But there is one country which dominates the new mining in Latin America far beyond all others: Canada.

Canada today is the mining capital of the world, home to almost half of all publicly listed mining and exploration companies, according to Canadian government figures (Government of Canada, 2022). More mining companies are listed on the Toronto Stock Exchange (TSX) and the TSX Venture Exchange (TSXV) than on any other stock market in the world, and far more equity capital is raised in Canada for mining exploration and development than anywhere else (Richer La Flèche, 2022). Many of these companies hold no mining claims in Canada and no real connection with the country, but choose to register there because it offers several advantages over other jurisdictions.

Firstly, Canada provides a light-touch regulatory environment in which to raise capital, with ambiguous rules and weak oversight. Secondly, it provides substantial state support for the sector – from tax relief for companies and incentives for investors, to political, financial, and diplomatic support for mining operations abroad. Finally, it has largely sheltered its mining industry from any legal consequences related to its business overseas.

A good example of Canada's permissiveness towards its mining industry are the weak requirements on disclosure. Companies are only required to disclose risks if they consider there may be a negative effect on the value of their stock. In other words, any social and environmental issues arising from the company's activity are relevant only insofar as they reduce value for investors (Deneault and Sacher, 2012, p. 29). This means that even when companies may be implicated in serious environmental and human rights abuses, it is easy for them to keep their investors in the dark and continue to do business in Canada.

It's no coincidence that Canada is the jurisdiction of choice for junior exploration companies: firms which focus on locating and developing mineral deposits. Most have neither the resources nor the experience required to actually operate a mine and will never extract any metal. Their aim is simply to sell on their concessions for a profit, sometimes retaining a minority share in the hope of receiving some returns if the project ever makes it to the extraction stage (PODER et al., 2017, p. 57).

The success of juniors is based purely on the market hype they can generate and the lax requirements on disclosure help them do just this – exaggerating the potential of their concessions on the one hand, while downplaying or simply omitting any problems on the other. The example of Almaden Minerals (see Chapter 3) shows that even when the value of a mining claim has clearly been compromised – for instance, by problems with licensing, or obtaining consent for operations from local communities – companies may misrepresent such information to their shareholders or simply fail to disclose.

This matters for investors, who may be duped into putting their capital into ventures which are much riskier than they seem at face value. But most of all, it matters for communities and ecosystems around the world, which are left with the negative 'externalities' of mining – the costs of company activity not listed on balance sheets or recognized in public statements. These externalities may be long term or irreversible.

Not America's backyard, but Canada's neighbourhood

In Latin America, there were just two operational mines in Canadian hands in 1990; by 2012, that had leapt to 80 (Gordon and Webber, 2016, p. 17). It is currently estimated that between 50 and 70 per cent of all mining activity in Latin America involves companies listed in Canada (Lama, 2021), ranging from giants like Barrick Gold, Yamana Gold, and Pan American Silver, all of which operate several mines across the region, to juniors which may only hold a single claim.

This dramatic influx of the Canadian mining industry into Latin America is no accident of globalization: it is the result of decisions made at the very highest levels of the Canadian state since the late 1980s by both Liberal and Conservative governments. Mining and hydrocarbon extraction are inherently risky industries; the goal of Canadian policy in Latin America, therefore, has been to protect Canadian interests and reduce this risk to a minimum (Gordon and Webber, 2016, p. 3).

Embassies across the region have acted on behalf of the mining industry, lobbying local politicians, and gathering intelligence to report back to industry leaders and policymakers in Ottawa. Export Development Canada – a Crown corporation owned by the Canadian state – provides mining companies with loans and insurance, guaranteeing companies against 'political risks' such as expropriation.

Canadian development aid has also been increasingly tied to projects which facilitate Canada's commercial interests, particularly mining. In 2011, for example, the Canadian government announced that the Canadian International Development Agency (CIDA) was to co-fund development projects in Africa and Latin America with the mining companies IAMGOLD, Rio Tinto Alcan, and Barrick Gold (Roth, 2017, p. 11). Essentially, this was the Canadian state subsidizing the Corporate Social Responsibility (CSR) programmes of three of Canada's richest mining companies, helping them neutralize local opposition to their projects and advance their interests further (see Chapter 3 for more on CSR, including that of Barrick). At the same time, CIDA cut funding to NGOs and charities in Latin America which had worked with groups critical of Canadian mining (Gordon and Webber, 2016, pp. 25–26).

A spokesperson for the then prime minister Stephen Harper was very candid about this strategy: 'Our government is strengthening its development assistance in the Americas because this is our neighbourhood, where we have significant interests' (ibid., p. 25).

In 2013, in a further indication that Canada was tying its development aid to its commercial interests, CIDA was abolished and its functions merged with the Department of Foreign Affairs and International Trade. Since then, the Canadian government has continued to provide funding for CSR projects in mining areas across the Global South (Roth, 2017, p. 12).

Canada's recent foreign policy objectives in Latin America have also been defined to a large extent by its commercial interests, particularly mining. This involved support for Álvaro Uribe's 'democratic security' agenda in Colombia, under which large swathes of the countryside were militarized – resulting in mass displacement and human rights abuses – partly in order to facilitate the expansion of the mining and energy sectors. Likewise, in 2008, the Canadian embassy in Ecuador and Canadian mining executives aggressively lobbied Rafael Correa's government during the drafting of a new mining law, resulting in legislation broadly favourable to the industry.

But the most flagrant example of Canada's pro-mining foreign policy lies in its intervention in the small Central American nation of Honduras, following a military coup in 2009 which removed the democratically elected president Manuel Zelaya (see Chapter 4). Along with the U.S. State Department under Hillary Clinton, Canada was one of the principal supporters of the authoritarian regime that seized power. The Harper government failed to condemn the brutal repression unleashed on pro-democracy protestors in the wake of the coup, while repeatedly insinuating that Zelaya and his supporters bore some responsibility for the situation of instability and violence (Gordon and Webber, 2016, p. 50).

In the following years, Canada worked hard to rehabilitate the isolated Honduran regime on the international stage, while seeking opportunities for Canadian business interests in the country. In 2011, Harper went to Honduras to sign an initial version of the Canada–Honduras Free Trade Agreement, becoming the first foreign leader to visit after the coup. Since then, Canadian investment has flooded into the country, particularly in the *maquilas* (poorly regulated, low-wage manufacturing), tourist developments, and of course, mining (ibid, p. 65).

Anti-mining movements in Honduras had secured a *de facto* moratorium on new concessions in 2004, ratified by Zelaya in 2006 amidst growing concern about the industry's environmental impacts. But with Zelaya in exile, Canada's mining industry seized its chance, assisted by the Canadian embassy, which brokered meetings between Canadian executives and leading politicians in the new regime. This resulted in a new mining law in 2013, which established very attractive conditions for Canadian companies and investors, but was widely denounced by critics of the industry in Honduras.

The result was a surge in social conflict in the countryside which, in the repressive context of post-coup Honduras, meant human rights abuses: from harassment, threats, and intimidation; to activists being criminalized and sent to jail; to serious injury and murder.

Shifting precedents

In 2016, a report by the Justice and Accountability Project found that there had been 44 deaths associated with the presence of Canadian mining companies in Latin America from 2000 to 2015, 30 of which were targeted assassinations. It also documented hundreds of injuries and cases of criminalization. All these abuses were widespread, occurring in at least eleven countries throughout the region. The authors stressed that the incidents they documented appeared to be merely the tip of the iceberg, with many other reports they were unable to corroborate (Imai et al., pp. 4–5).

Both the United Nations and the Inter-American Commission on Human Rights have called on Canada to hold its mining industry accountable for this catalogue of abuses, but the Canadian state has largely sat on its hands (Kuyek, 2019, p. 128).

In 2009, during Stephen Harper's minority Conservative government, the Liberal MP John McKay tabled Bill C-300, or the *Corporate Accountability of Mining, Oil and Gas Corporations in Developing Countries Act*. This would have made state support for Canadian mining companies operating abroad – for example, access to credit from Export Development Canada – contingent on compliance with international human rights and environmental standards.

Though a step in the right direction, critics argued that the bill did not go far enough. Rather than establishing a dedicated, independent watchdog to evaluate allegations of abuse against Canadian companies, the bill assigned the responsibility to government ministers, who would have been susceptible to political pressure. Moreover, these ministers would only have been able to issue 'guidelines' to companies – not binding legal regulations.

Even so, Canada's mining industry mobilized aggressively against the bill, claiming it would damage the competitiveness of Canada's mining sector and force companies to relocate. In the end, it was narrowly defeated in the House of Commons by the Conservatives, assisted by 13 Liberal MPs who failed to show up for the vote. Two related bills, tabled between 2010 and 2013, failed to progress beyond the initial stages.

This reluctance to rein in the country's mining sector is not limited to the Conservatives. During the election campaign of 2015, Justin Trudeau's Liberals pledged to create a new watchdog to investigate claims of human rights abuses involving Canadian companies overseas. Crucially, it was to be given the power to force company employees to testify and hand over documents.

This watchdog – known as the Canadian Ombudsperson for Responsible Enterprise (CORE) – was eventually created in January 2018. But intense industry lobbying in the interim managed to remove most of its teeth. It now has no power to compel witnesses or documents, meaning its investigations depend on the voluntary cooperation of the companies accused. And while the CORE may advise companies and government on how to rectify any harm done, it has no power to enforce any of its recommendations (Pazzano, 2021).

In the absence of any serious regulation, it has been left to the Canadian courts to intervene. In the past, companies have managed to dodge any legal consequences in Canada for abuses abroad by invoking a principle known as *forum non conveniens*; in other words, they argue the Canadian courts are not the correct place for the case to be heard. Another common defence is that the parent company in Canada is not responsible for abuses committed by subsidiaries located overseas. Cases have tended to return to plaintiffs' home countries, where they usually dwindle and die out, given the weakness of local court systems.

But this may be changing. Since 2013, several cases against Canadian companies have been successfully brought before Canadian courts. Three were brought by indigenous Maya Q'eqchi plaintiffs from El Estor, Guatemala, who accuse Hudbay Minerals and its predecessor Skye Resources of gross human

Photo 2.2 Q'eqchi plaintiffs from El Estor, Guatemala / © James Rodríguez 2014

rights violations at the Fenix nickel mine in 2007 and 2009. These alleged abuses consist of the brutal murder of a community leader; the shooting of a young farmer, resulting in his paralysis; and the gang rape of eleven women by soldiers, police officers, and security guards. As of early 2022, the cases remain ongoing.

In a separate case, a Canadian judge ruled in 2017 that a case brought by seven Guatemalan plaintiffs against Tahoe Resources could proceed in Canada. The plaintiffs were anti-mining demonstrators who had been shot and injured outside the controversial Escobal mine by Tahoe's security personnel in 2013, in what they alleged was an operation planned by Tahoe's head of security and approved by the company. In 2019 Pan American Silver bought out Tahoe and negotiated a settlement, issuing a public apology for the actions of its predecessor (Global Freedom of Expression, 2022).

Then, in a landmark ruling in February 2020, Canada's Supreme Court ruled that a Vancouver-based mining company could be sued in Canada over its alleged use of slave labour in Eritrea. This is the strongest indication yet that companies will no longer be able to use Canadian jurisdiction as a shield against allegations of human rights abuses committed abroad. In theory, this could lead to a wave of civil cases brought by Latin American plaintiffs against Canadian mining companies.

***

Canada has long enjoyed something of a glowing reputation on the international stage: peaceful, progressive, internationalist, a safe harbour for immigrants and refugees from all over the world. But its recent policies in

Latin America and elsewhere in the Global South – particularly where mining is concerned – have done much to tarnish this image.

Across Latin America, the Canadian state has systematically intervened in local affairs, in support of an industry much compromised by allegations of environmental destruction and human rights abuses. At the same time, successive Canadian governments have resisted any serious attempt to rein in the mining sector or hold it accountable for these excesses. As a result, a growing body of opinion argues that Canada has become, in the words of Noam Chomsky, 'a secondary imperialist power' (Gordon and Webber, 2016, dust jacket).

To illustrate this in more detail, we now present a notorious case of Canadian corporate malfeasance in Latin America, involving the use of a controversial legal system which has enabled companies based in Canada and elsewhere to extract billions of dollars from Latin America – and sometimes without even getting a drill in the ground.

IV. ISDS: making a killing

On 3 June 2011, Juan Francisco Durán Ayala, a 30-year-old postgraduate student and environmental activist, left home to attend class at the Technological University of San Salvador, El Salvador's sprawling capital city. It was the last time he was seen alive. Shortly after midnight on 4 June, his body was found. He had been shot twice in the head.

Declaring his body 'unidentifiable', the authorities simply buried Durán in a common grave. It wasn't for another ten days that Durán's allies from the Environmental Committee of Cabañas (CAC) found out what had happened. His father positively identified his body that same day.

The department of Cabañas is a hostile environment for activists. Staunchly conservative, it is a stronghold of the military, which terrorized the civilian population during El Salvador's twelve-year long civil war (1979–92). From the early 2000s, it became the site of a fresh conflict; this time, between locals and the Canadian mining company Pacific Rim, which was attempting to install a goldmine near the town of San Isidro. It was called El Dorado, recalling the mythical city of gold which had so obsessed the Spanish *conquistadores* and other European explorers.

The day before his disappearance, Durán had been seen in the town of Ilobasco, just west of San Isidro, putting up posters and handing out flyers as part of the campaign against Pacific Rim. According to the CAC, the local mayor had ordered the police to remove the posters and intimidate anyone seen putting them up.

Durán was the last of at least five people who appear to have been killed in retaliation for opposition to Pacific Rim. The other four killings all took place in 2009. In June that year, Marcelo Rivera, a local artist and community organiser, was found at the bottom of a well; his fingernails had been pulled out and his trachea broken. In December, Ramiro Rivera (no relation) died when gunmen armed with M-16 assault rifles riddled his pickup truck with bullets. His partner, Felicita Echeverría, was also killed. Six days later, Dora Alicia Recinos was shot dead as she returned from the spring where she had been washing clothes. She was eight months pregnant at the time.

Many other activists and journalists in Cabañas also reported harassment and threats in relation to their resistance to Pacific Rim. In particular, reporters at the community radio station Radio Victoria were repeatedly targeted for covering Pacific Rim's activities and the destructive impacts of industrial mining.

'Even during the war I wasn't so afraid of being killed,' said Hector Berrios, another activist and a friend of Marcelo Rivera (Provost, 2014a). 'I would get calls in the middle of the night saying they know where my daughter goes to school, where my wife is, and that they're going to assassinate her if I don't leave town' (Nugent, 2015).

Affectionately known as *el pulgarcito de América* – 'the Tom Thumb of the Americas' – El Salvador is a tiny country of around just 21,000 km^2, only slightly bigger than Wales. But it consistently registers some of the highest murder rates in the world.

During the 1990s, the United States began mass deportations of young, foreign-born men with criminal records, many of whom had been involved in Latino gangs in American cities, particularly Los Angeles. When these deportees arrived back in El Salvador, they found themselves in a country struggling to overcome the devastation of war, able to provide little in terms of employment or opportunity. This sudden influx of hardened gang members into a weak, war-torn state was a recipe for disaster. Within a decade, a brutal three-way conflict between the Mara Salvatrucha and Barrio 18 gangs and the Salvadoran security forces had erupted, and it has been raging ever since.

Currently, there are around 60,000 gang members active in the country, who are notorious for their cruelty; torture, dismemberment, and sexual violence are not uncommon. Their influence permeates everyday life: neighbourhoods, schools, public transport, the police. These atrocious levels of violence have made it easy for the police and local authorities to depoliticize the killings of mining opponents, writing them off as the work of the gangs. Nonetheless, the Salvadoran Ombudsman for Human Rights stated that the killings in Cabañas 'are very probably related to each other, thus enabling us to infer that they are also linked to the victims' work in defense of the environment' (Moore et al., 2014, p. 9).

Pacific Rim has always denied any involvement in violence towards activists in Cabañas. But at the very least, the presence of the company heightened tensions in the area, creating bitter divisions between supporters and opponents of the company. Pacific Rim sought alliances with local conservative politicians, which quickly generated conflict with activists, priests, and journalists who had spoken out against mining (ibid., p. 8).

And in El Salvador, all too often such conflicts are settled at gunpoint.

'No winners'

The story of Pacific Rim has become an emblematic case of mining conflict not only because of these human rights abuses, but also because it threw the spotlight on a secretive and controversial legal system known as investor-state-dispute-settlement (ISDS).

Most of Latin America and the Caribbean became entangled with ISDS during the neoliberal reforms of the 1990s. In the hope of attracting foreign investment, governments all over the region signed hundreds of bilateral investment treaties and free trade agreements which granted unprecedented rights to foreign investors. Among these rights was ISDS, which allows investors to sue nation states when they believe that decisions taken by their governments or courts may have had a negative impact on the value of their investment.

The cases are heard before international arbitration tribunals, such as the International Centre for Settlement of Investment Disputes (ICSID) at the World Bank, or the United Nations Commission on International Trade Law (UNCITRAL). These tribunals have a built-in bias towards companies, being overseen not by independent judges, but three-person panels of corporate lawyers who meet behind closed doors. Lawyers may also switch roles, sitting as an arbitrator on one case, but acting as counsel for a corporation on another. It is a highly lucrative industry: these lawyers may charge as much as $1,000 an hour for their services and so law firms lobby corporations to sue whenever possible (Leadnow.ca, 2014).

Lawsuits may also be funded by third parties – usually a hedge fund or finance firm – in return for a slice of the rewards in the event of a favourable ruling. Such funders are not liable for any costs if the case is a failure, which has encouraged speculative high-cost claims which have little to do with restitution and far more to do with making an easy profit (Smith, 2018, p. 33). Stakes in ISDS claims can even be treated as derivatives, to be bought and sold like any other financial product. For one leading arbitrator, third-party funding of ISDS is 'a gambler's Nirvana' (ibid., p. 32).

Pacific Rim filed against El Salvador at ICSID in 2009 and the case proceeded under ISDS provisions in El Salvador's own national investment law. Having submitted a final design for El Dorado in October 2006, the company was frustrated it hadn't been granted permission to begin mining. But this was due to a *de facto* moratorium on new mining projects, introduced by the right-wing ARENA government in 2006 amidst growing public concern about the impacts of industrial mining, particularly on water supply. El Salvador is one of the most environmentally degraded and water-stressed countries in the Americas; at least 90 per cent of the country's surface water has been contaminated by sewage, industrial wastewater, and chemicals from agriculture (Lakhani, 2019). Opposition to mining had also been galvanized by evidence of pollution of the San Sebastián river in the east of the country, caused by acid mine drainage from an American-owned goldmine.

Photo 2.3 Children bathing in the Río Lempa, El Salvador's longest and only navigable river. Activists highlighted the threat posed to the river by mining / kadejo 2007 / CC BY-SA 3.0

ISDS permits companies to sue not only for the money they have spent attempting to develop a concession, but also for projected losses. Pacific Rim, for example, sued El Salvador for $314 million – far more than the $77 million it claimed to have spent – based on the hypothetical profits it expected to make had it been given authorization to mine. This was a significant sum for El Salvador, equivalent to around two per cent of national GDP (Nugent, 2015), and more than it received annually in aid in any single year from 2011 to 2019 (OECD, 2022).

Pacific Rim nearly went bankrupt in 2013 and was bought out by the Canadian-Australian firm OceanaGold for just $10.2 million – many times less than the value of the lawsuit. With the spurious justification that it wished 'to gain clarity on the Government of El Salvador's position on the permit application filed by Pacific Rim', OceanaGold chose to press ahead with the case (OceanaGold, 2022).

In its defence, El Salvador's legal team argued that Pacific Rim/OceanaGold had failed to submit a feasibility study; its Environmental Impact Assessment (EIA) had never been approved; and the company had failed even to prove it had obtained the rights to the land it needed to develop the mine, given that many local landowners had refused either to sell up or to grant the company permission to work on their property (Provost, 2014b).

In 2016, the ICSID tribunal reached its verdict. It ruled unanimously in favour of El Salvador, finding that Pacific Rim/OceanaGold had failed to meet the criteria laid out by the national mining law and that El Salvador was within its rights to deny permission to mine.

Had justice been done? Not exactly.

The case had dragged on for seven years, draining time, energy, and resources away from a country with other major concerns. While the tribunal ordered Pacific Rim/OceanaGold to reimburse El Salvador $8 million for its legal fees, the country had spent close to $13 million on its defence, and of course, it received no compensation at all for the impacts of Pacific Rim's exploration work. Finally, the case had been the backdrop to a conflict that claimed the lives of at least five people.

The NGO MiningWatch Canada, which had closely monitored the case, was blunt in its assessment: there were 'no winners' (International Allies Against Mining in El Salvador, 2016).

A new El Dorado

ISDS is a recent phenomenon which only began to grow significantly after the turn of the millennium. From just six recorded ISDS cases globally in 1995, there were more than 1,100 as of late 2020. Latin America and the Caribbean are disproportionately affected, accounting for 303, or 27.4 per cent. Of these, 23.1 per cent concern mining and hydrocarbons, more than any other sector (ISDS America Latina, 2021).

In terms of outcome, the Pacific Rim case was exceptional. Though they are not always awarded the full value of the claim, investors tend not to end up empty handed – whether because a tribunal rules in their favour, or because they reach a settlement with the government in question. And given that nearly 87 per cent of the lawsuits are filed by Americans, Canadians, and Europeans, for investors from the Global North, this system has itself become a new El Dorado (ibid.).

Pacific Rim v. El Salvador is the most well-known case involving a mining company in Latin America, but there have been at least 34 other such lawsuits since 1997 (Boorer, 2021). One notorious example comes from Ecuador, where in 2008 the government revoked concessions belonging to Canadian company Copper Mesa Mining, arguing it had failed to produce a valid EIA or adequately consult local communities.

The concession was located in the Intag Valley, a mountainous region of extreme ecological sensitivity in the north of the country. Local opposition had been fierce from the beginning, essentially preventing Copper Mesa from getting any serious exploration work done (Moore and Pérez-Rocha, 2019, p. 23). In response, the company resorted to threats, intimidation, and violence. This included legal action against mining opponents, such as a million-dollar libel suit against a community newspaper (it was eventually dropped), and the use of firearms and tear gas against the local population (Fieweger, 2005).

Yet Copper Mesa took Ecuador to UNCITRAL and won, with arbitrators ruling that due process had not been followed. The arbitrators claimed that

Photo 2.4 Guards from the Ecuadorian security company Securivital attacking villagers from Junín, in the Intag Valley / © Elisabeth Weydt 2006

Ecuador 'should have attempted something to assist' Copper Mesa get the project off the ground, despite also recognizing that the company's tactics had included 'recruiting and using armed men, firing guns and spraying mace at civilians, not as an accidental or isolated incident but as part of a premeditated, disguised and well-funded plan to take the law into its own hands' (Beachy, 2016).

Few countries saw as radical a shift towards mining during this period as Colombia. Under president Álvaro Uribe (2002–10), the area of national territory under mining concession increased eightfold (Armeni, 2011), a process so haphazard it was labelled 'the *piñata* of mining concessions' by Carlos Rodado, minister of mines and energy under Uribe's successor Juan Manuel Santos. Many of these new concessions overlapped with indigenous lands and other sensitive areas and some investors have faced major social and legal obstacles to developing them. This has led to a surge in ISDS cases (Chacón González, 2011).

A good example from this '*piñata*' was the granting of a gold concession deep in the Amazon rainforest to the Canadian company Cosigo Resources in 2009, though a national park precluding any mining had been created over the land in question just two days beforehand. Despite having done little exploration work of any substance, and having faced considerable resistance from indigenous communities in the region, Cosigo is suing Colombia at UNCITRAL for the outrageous sum of $16.5 billion (Moore and Pérez-Rocha, 2019; Castro, 2017). As of May 2022, the case remains pending.

In theory, ISDS mechanisms were created to encourage foreign investment by reducing risk for investors. But the reality has proven very different. Companies and investors are using ISDS to bully nation states into acting in their interests. Cosigo, for example, has promised to drop its exorbitant claim if Colombia will do something to stop the national park from blocking its plans. Furthermore, ISDS has become a means of blatant financial speculation. Many of these cases are brought by exploration companies which have no operating mine and no realistic prospect of developing one, but they sue anyway, hoping to squeeze governments for whatever they can get (Moore and Pérez-Rocha, 2019, p. 16). Their chances are reasonable, and third-party funding means that even if the lawsuit fails, the company may not have to pick up the tab.

'I think the investor-state arbitration system was created with good intentions, but in practice it has gone completely rogue,' says Luis Parada, a lawyer who defended El Salvador in the Pacific Rim case (Provost and Kennard, 2015).

End of a bonanza?

In most of the ISDS cases brought by mining companies against Latin American states, some degree of community resistance has been involved. But communities are shut out of the tribunals and their voices go mostly unheard by arbitrators.

Nonetheless, these cases do not take place in a vacuum. While officially El Salvador's victory may have been on technical grounds, it also took place against the backdrop of a well-orchestrated campaign against Pacific Rim at the local, national, and even international level. There was global media coverage of the case, and protests against Pacific Rim in the United States, Canada, and Australia. This all helped pile the pressure on the lawyers who sat on the arbitration tribunal.

It also propelled a nationwide movement for a total ban on metal mining in El Salvador, led by The National Roundtable Against Metal Mining, better known as La Mesa, which brought together affected communities, thinktanks, NGOs, academics, and religious organizations. In 2017, in a watershed moment for anti-mining movements across Latin America and beyond, Salvadoran lawmakers voted unanimously in favour of the ban, making El Salvador the first – and to date, the only – country ever to do so.

In late 2021, La Mesa denounced a visit to El Salvador by the Intergovernmental Forum on Mining, Metals and Sustainable Development, an organization with strong Canadian presence and core funding from the Canadian government (ARPAS, 2021). Though the stated aim was to help El Salvador develop other forms of mining, critics fear this is a smokescreen. Moreover, with President Nayib Bukele rapidly dismantling the country's democratic institutions and shrinking civic space, there are concerns that the mining ban could be overturned.

ISDS has been a disaster for Latin America. It has siphoned billions of dollars from national budgets into the pockets of investors located overwhelmingly in the Global North. The spectre of expensive and time-consuming ISDS cases limits the ability of governments and court systems to act in the public interest. And in some cases, such as *Pacific Rim v. El Salvador*, ISDS has exacerbated tensions on the ground and made conditions more dangerous for environmental and human rights defenders.

As a result, several countries have been attempting to disengage from the system. Bolivia's 2009 Constitution stipulates that all investment disputes must be resolved under national jurisdiction. Ecuador, which has been hit extremely hard by ISDS cases brought by oil companies, terminated all of its bilateral investment treaties under President Rafael Correa (2007–17), though his successors have reversed this policy. And in 2013, El Salvador modified its investment law – originally passed in 1999 with input from the World Bank – to ensure that cases were first heard in national courts, before going to international arbitration.

But for Parada, reforming the ISDS system is easier said than done. 'The states that created the system are the only ones that can fix it,' he says. 'I have not seen a critical mass of states with the political will, much less a broad consensus. But I still hope it happens' (Provost and Kennard, 2015).

***

Mining has been a constant in Latin America since colonial times, shaping the region's economic development and social geography like almost no other industry. The search for precious metals defined early relations between indigenous people and the Spanish. It was a key motor of the slave trade, with Africans being shipped across the Atlantic to work in the silver mines of Spanish America and the goldfields of Brazil. It provoked mass movements of people hoping to make their fortune – both from Europe to Latin America, and from the coastal regions to the interior. Finally, it established Latin America as a provider of raw materials to global markets.

However, since the early 1990s, mining in Latin America has metamorphosed, taking on shapes and dimensions which are fundamentally different to earlier periods, and with the participation of new global actors, most notably Canada (Bebbington et al., 2013, p. 267). This process has been fuelled by high demand for the abundant metals and minerals found in the region's subsoil, and today, prices look set to rise further still, given the need to transition away from fossil fuels and towards renewable energy and green technologies.

Yet despite unprecedented new flows of foreign investment, and the extraordinary profits generated for companies and investors during this period, mining has comprehensively failed to deliver on its promise of prosperity for the people of Latin America.

This is particularly true for the communities located closest to operations. Such communities are sometimes poor, remote, and may have longstanding social problems which governments have shown little interest in addressing.

But not only does mining provide no credible long-term solution to these existing problems, it leaves behind a whole host of others: water scarcity; pollution of the air, water, and soil; the threat of tailings dam failures; the division of communities; and the destruction of traditional cultures and livelihoods, among others.

But while the forces stacked against them may seem at once distant and overwhelming, communities are not simply passive bystanders in all this. Community resistance is often the key to rejecting mining projects at the local level and has succeeded in changing the law at the provincial and even national level, as in the case of Pacific Rim. That is precisely why companies put considerable resources into CSR and community engagement, as we will see in the next chapter.

Note

1. According to the National Survey of Living Conditions (ENCOVI) by researchers at the Andres Bello Catholic University, Venezuela's ongoing economic crisis has seen poverty rates surge to the highest in Latin America.

References

All references to web-based material were checked and still available in November 2022 unless otherwise stated.

All references are listed, with clickable links for your convenience, on the page for this chapter on the Heart of Our Earth website: <https://lab.org.uk/the-heart-of-our-earth/>

Armeni, A. (2011) 'Mining: The Risks for Afro-Colombians and the Indigenous'. [online] *Americas Quarterly*. Available at: <https://www.americasquarterly.org/fulltextarticle/mining-the-risks-for-afro-colombians-and-the-indigenous/>.

ARPAS (2021) 'Advierten intención del gobierno de permitir minería en el país'. [online] Available at: <https://arpas.org.sv/2021/12/advierten-intencion-del-gobierno-de-permitir-mineria-en-el-pais/>.

Balch, O. (2019) 'How a populist president helped Bolivia's poor – but built himself a palace'. [online] *The Guardian*. Available at: <https://www.theguardian.com/world/2019/mar/07/how-a-populist-president-helped-bolivias-poor-but-built-himself-a-palace>.

Ballard, K., Cline, R., Neubig, T. and Phillips, A. (2012) 'Mining Taxation and Global Investment: Evaluating Trade-Offs Between "Fair Share" of Resource Rents and Mining Development'. *Proceedings. Annual Conference on Taxation and Minutes of the Annual Meeting of the National Tax Association, 105th Annual Conference on Taxation (November 15–17, 2012)*, [online] pp. 55–63. Available at: <https://www.jstor.org/stable/prancotamamnta.105.55>.

Beachy, B. (2016) 'Why Mining Corporations Love Trade Deals'. [online] Common Dreams. Available at: <https://www.commondreams.org/views/2016/06/22/why-mining-corporations-love-trade-deals>.

Bebbington, A. (2010) 'Extractive Industries and Stunted States: Conflict, Responsibility and Institutional Change in the Andes'. In: K. Raman and R. Lipschutz, eds., *Corporate Social Responsibility: Comparative Critiques.* [online] Basingstoke, UK: Palgrave Macmillan, pp. 97–115. Available at: <https://wordpress.clarku.edu/abebbington/files/2010/12/Chapter-2010-Extractive-industries-and-stunted-states.pdf>.

Bebbington, A. and Bury, J. (2013) 'Political ecologies of the subsoil'. In: A. Bebbington and J. Bury, eds., *Subterranean Struggles. New dynamics of mining, oil, and gas in Latin America.* Austin, TX: University of Texas Press, pp. 1–26.

Bebbington, A., Bury, J. and Gallagher, E. (2013) 'Conclusions'. In: A. Bebbington and J. Bury, eds., *Subterranean Struggles. New dynamics of mining, oil, and gas in Latin America.* Austin, TX: University of Texas Press, pp. 267–288.

Boorer, M. (2021) 'Mining: ISDS – a licence to plunder'. [online] Latin America Bureau. Available at: <https://lab.org.uk/mining-isds-a-licence-to-plunder/>.

Britannica (n.d.) 'Bolivia - Increase in tin mining'. [online] Available at: <https://www.britannica.com/place/Bolivia/Increase-in-tin-mining>.

Brown, K. (2012) *A History of Mining in Latin America. From the Colonial Era to the Present.* Albuquerque: University of New Mexico Press.

Bury, J. and Bebbington, A. (2013) 'New Geographies of Extractive Industries in Latin America'. In: A. Bebbington and J. Bury, eds., *Subterranean Struggles. New dynamics of mining, oil, and gas in Latin America.* Austin, TX: University of Texas Press, pp. 27–66.

Castro, C. (2017) 'La minera que pide 16.500 millones de dólares de indemnización a Colombia'. [online] *Semana.* Available at: <https://www.semana.com/nacion/articulo/16500-millones-de-dolares-demanda-tobie-mining-contra-colombia-por-apaporis/528264/>.

Chacón González, J. (2011) 'La piñata de los títulos mineros'. [online] *El Espectador.* Available at: <https://www.elespectador.com/economia/la-pinata-de-los-titulos-mineros-article-273872/>.

Deneault, A. and Sacher, W. (2012) *Imperial Canada Inc. Legal Haven of Choice for the World's Mining Industries.* Vancouver: Talonbooks.

Fieweger, M. (2005) 'Mining Transnational asking one million dollars in damages'. [online] Mines and Communities. Available at: <http://www.minesandcommunities.org/article.php?a=342uihn>.

Francescone, K. (2021) 'El daño en el cerro está hecho'. [online] *Página Siete.* Available at: <https://www.paginasiete.bo/economia/2021/12/13/una-docena-de-hundimientos-amenazan-el-cerro-rico-tres-de-ellos-en-la-cima-317933.html#!>.

Global Freedom of Expression (2022) 'García v. Tahoe Resources Inc.'. [online] Available at: <https://globalfreedomofexpression.columbia.edu/cases/garcia-v-tahoe-resources-inc/>.

Gordon, T. and Webber, J. (2016) *Blood of Extraction. Canadian imperialism in Latin America.* Halifax, Nova Scotia and Winnipeg, Manitoba: Fernwood Publishing.

Government of Canada (2022) 'Canadian Mining Assets'. [online] Available at: <https://www.nrcan.gc.ca/maps-tools-and-publications/publications/minerals-mining-publications/canadian-mining-assets/19323>.

Greenfield, P. (2016) 'Story of cities #6: how silver turned Potosí into 'the first city of capitalism''. [online] *The Guardian*. Available at: <https://www.theguardian.com/cities/2016/mar/21/story-of-cities-6-potosi-bolivia-peru-inca-first-city-capitalism>.

Grieco, K. (2016) 'Motherhood, Mining and Modernity in the Peruvian Highlands from Corporate Development to Social Mobilization'. In: N. Dhawan, E. Fink, J. Leinius and R. Mageza-Barthel, eds., *Negotiating Normativity*. Switzerland: Springer International Publishing, pp. 131–146.

Grupo de Trabajo sobre Minería y Derechos Humanos en América Latina (2014) *El impacto de la minería canadiense en América Latina y la responsabilidad de Canadá*. [online] Available at: <https://www.dplf.org/sites/default/files/informe_canada_resumen_ejecutivo.pdf>.

Harris, P. (2020) 'Cerro Rico production stops after nearly 500 years'. [online] *Mining Journal*. Available at: <https://www.mining-journal.com/covid-19/news/1385532/cerro-rico-production-stops-after-nearly-500-years>.

Imai, S., Gardner, L. and Weinberger, S. (2017) 'The 'Canada Brand': Violence and Canadian Mining Companies in Latin America'. *Osgoode Legal Studies Research Paper No. 17/2017*, [online] Available at: <https://papers.ssrn.com/sol3/Delivery.cfm/SSRN_ID3082286_code1019085.pdf?abstractid=2886584&mirid=1>.

Instituto Nacional de Estadística (2022) 'Bolivia: incidencia de pobreza extrema, según departamento, 2011–2018'. [online] Available at: <https://nube.ine.gob.bo/index.php/s/P8eMMLJbbzMWaUV/download>.

International Allies Against Mining in El Salvador (2016) 'There Are No Winners: After Seven Years and Millions of Dollars, Pac Rim Mining Loses Suit Against El Salvador'. [online] MiningWatch Canada. Available at: <https://miningwatch.ca/news/2016/10/14/there-are-no-winners-after-seven-years-and-millions-dollars-pac-rim-mining-loses>.

ISDS America Latina (2021) 'En numeros - ISDS America Latina'. [online] Available at: <https://isds-americalatina.org/en-numeros/>.

Kuyek, J. (2019) *Unearthing Justice. How to protect your community from the mining industry*. Toronto: Between The Lines.

Lakhani, N. (2019) 'Living without water: the crisis pushing people out of El Salvador'. [online] *The Guardian*. Available at: <https://www.theguardian.com/global-development/2019/jul/30/el-salvador-water-crisis-privatization-gangs-corruption>.

Lama, C. (2021) 'A Dark Side to Canada: An Analysis of Canadian Mining Corporations in Latin America'. [online] *The Observer*. Available at: <https://theobserver-qiaa.org/a-dark-side-to-canada-an-analysis-of-canadian-mining-corporations-in-latin-america>.

Leadnow.ca (2014) *What is an Investor-State Dispute Settlement clause?*. [video] Available at: <https://www.youtube.com/watch?v=2SbO2zDDpDA>.

Machado, R. and Fachin, P. (2015) 'O modelo neoextrativista e o paradoxo latino-americano. Entrevista especial com Bruno Milanez'. *Cadernos IHU em formação. Mineração e o impulso à desigualdade: impactos ambientais e sociais*, [online] 48, pp. 60–64. Available at: <http://www.ihu.unisinos.br/images/stories/cadernos/formacao/48ihuemformacao.pdf>.

MiningWatch Canada (2005) 'The Taxman Needs to Help Shift from Mining Virgin Materials to Recycling'. [online] Available at: <https://miningwatch.ca/fr/node/5247>.

MiningWatch Canada (2021) 'Bolivian Citizens Block Mill Operations, Accuse Canadian Company of Destroying UNESCO World Heritage Site'. [online] Available at: <https://miningwatch.ca/news/2021/10/20/bolivian-citizens-block-mill-operations-accuse-canadian-company-destroying-unesco>.

Moore, J., Broad, R., Cavanagh, J., Guerra Salazar, R., Karunananthan, M., Morrill, J., Pérez-Rocha, M. and Vergara, S. (2014) *Debunking Eight Falsehoods by Pacific Rim Mining/OceanaGold in El Salvador*. [online] Available at: <https://issuu.com/pedrocabezas/docs/eight_falsehoods_final__march_17_20>.

Moore, J. and Pérez-Rocha, M. (2019) *Extraction Casino. Mining companies gambling with Latin American lives and sovereignty through supranational arbitration*. [online] Available at: <https://miningwatch.ca/sites/default/files/isds_report_final_0.pdf>.

Nugent, C. (2015) 'El Salvador vs Pacific Rim: the price of saying 'no' to a gold mine'. [online] bilaterals.org. Available at: <https://www.bilaterals.org/?el-salvador-vs-pacific-rim-the>.

OceanaGold (2022) 'El Dorado Project'. [online] Available at: <https://oceanagold.com/operation/closure/el-dorado-project-el-salvador/>

OECD (2022) 'Creditor Reporting System (CRS)'. [online] Available at: <https://stats.oecd.org/Index.aspx?DataSetCode=CRS1>.

Olave, M. (n.d.) *Recursos Naturales y Crecimiento Económico en Bolivia*. [online] Universidad Privada Boliviana. Available at: <https://www.upb.edu/sites/default/files/adjuntos/Paper%20Completo%20Recursos%20Naturales%20y%20crecimiento%20Económico%20en%20Bolivia.pdf>.

Pazzano, J. (2021) 'Trudeau government backpedals on investigating human rights complaints against mining companies'. [online] Global News. Available at: <https://globalnews.ca/news/7650556/human-rights-abuses-trudeau/>.

Petras, J. and Veltmeyer, H. (2018) 'Extractivism and resistance: a new era'. In: J. Petras and H. Veltmeyer, eds., *The Class Struggle in Latin America. Making History Today*. Abingdon, Oxon: Routledge, pp. 12–25.

PODER; Unión de Ejidos y Comunidades en Defensa de la Tierra, el Agua y la Vida, Atcolhua; CESDER; IMDEC (2017) *Minería canadiense en Puebla y su impacto en los derechos humanos*. [online] Available at: <https://poderlatam.org/wp-content/uploads/2020/02/InformeMineriaCanadiense-2017_SNP.pdf>.

Provost, C. (2014a) 'El Salvador's communities battle to keep their gold in the ground'. [online] *The Guardian*. Available at: <https://www.theguardian.com/global-development/2014/apr/11/el-salvador-communities-battle-gold-in-ground>.

Provost, C. (2014b) 'El Salvador groups accuse Pacific Rim of 'assault on democratic governance''. [online] *The Guardian*. Available at: <https://www.theguardian.com/global-development/2014/apr/10/el-salvador-pacific-rim-assault-democratic-governance>.

Provost, C. and Kennard, M. (2015) 'The obscure legal system that lets corporations sue countries'. [online] *The Guardian*. Available at: <https://www.theguardian.com/business/2015/jun/10/obscure-legal-system-lets-corportations-sue-states-ttip-icsid>.

Richer La Flèche, E. (2022) 'The Mining Law Review: Canada'. [online] The Law Reviews. Available at: <https://thelawreviews.co.uk/title/the-mining-law-review/canada_1>.

Roth, T. (2017) *Human Rights and the Canadian Extractive Sector in Latin America: Can Canada do more to prevent abuses and improve access to justice?* [online] University of Ottawa. Available at: <https://ruor.uottawa.ca/bitstream/10393/37123/1/ROTH,%20Tania%2020179.pdf>.

Servindi – Servicios de Comunicación Intercultural (2020) '"Los pasivos son la mejor evidencia del daño de la minería"'. [online] Available at: <https://www.servindi.org/actualidad-entrevistas-noticias/06/08/2020/los-pasivos-son-la-mejor-evidencia-del-dano-de-la-mineria>.

Smith, R. (2018) *"Tempering the Gambler's Nirvana". A Review into to the issues and regulation of Third Party Funding in Investment Treaty Arbitration.* Master's thesis. Uppsala Universitet. [online] Available at: <https://www.diva-portal.org/smash/get/diva2:1212686/FULLTEXT01.pdf>.

Tetreault, D. (2015) 'Social Environmental Mining Conflicts in Mexico'. *Latin American Perspectives*, 42(5), pp. 48–66.

CHAPTER 3
CSR: We're in charge now

This chapter examines Corporate Social Responsibility (CSR) and community engagement strategies, illustrating how companies have used these tactics to neutralize community resistance and create a favourable social environment for operations. It focuses first on efforts by the Canadian junior Almaden Minerals to obtain the so-called 'social licence to operate' for its Ixtaca gold–silver project in the Sierra Norte de Puebla, Mexico; and then on the conflict between the Canadian giant Barrick Gold and communities in the north of San Juan, Argentina, following repeated spills at its Veladero gold and silver mine.

I. Sierra Norte de Puebla, Mexico

For Morgan Poliquin, CEO of the Canadian mining company Almaden Minerals, unlocking the mineral potential of what he once called 'the Mexican unknown' has been his life's work (Poliquin, 2009, p. 2).

Poliquin has led extensive research and exploration work in eastern Mexico since at least 1994 (PODER et al., 2017, p. 58). During his PhD in earth resources at the University of Exeter, Poliquin travelled to Mexico on numerous occasions, surveying the region by helicopter during what he described in his thesis as 'many hours of dangerous flying' (Poliquin, 2009, p. 2). This exploration work was funded by Almaden and the Anglo-Australian mining giant BHP Billiton (now simply BHP). Poliquin was also backed by a substantial team of geologists and field staff and received support from the Pacific Centre for Isotopic and Geochemical Research at the University of British Columbia, as well as from Exeter's own Camborne School of Mines.

From the early 2000s, Poliquin increasingly began to focus his attention on the Sierra Norte de Puebla, a mountainous region roughly halfway between Mexico City and the Gulf Coast. Though just 140 km from the capital, the region has historically been isolated and it is only in recent decades that roads have been built and it has begun to develop. It's a dramatic landscape of steep valleys covered in dense forest, the varied elevation endowing it with rich biodiversity. There are spectacular waterfalls, as well as perhaps the most extensive network of caves in Latin America.

It is inhabited by the Nahua, Otomi, Tepehua, and Totonac indigenous peoples, as well as traditional rural mestizo communities (Ramos Mancilla, 2015, p. 13). Local producers grow a wide range of crops, from coffee, beans, and corn to potatoes, chilli peppers, and citrus fruit (Martínez et al., 2007, p. 18).

There are also a number of *ejidos* – plots of communal land recognized by the state, which are mostly used for forestry, small-scale livestock farming, and subsistence agriculture.

Fernanda Hopenhaym is a chief executive director at the Project on Organization, Development, Education and Research (PODER), a non-profit advocating for corporate accountability and transparency in Latin America. PODER has campaigned against megaprojects such as the new international airport which was to be built for Mexico City at Texcoco, and the Tren Maya railway on the Yucatán Peninsula, as well as working with mining-affected communities.

'The people are mainly small agricultural producers,' she says. 'They keep *milpas*[1] and need clean water for their crops. There are hills and woods; forestry has a lot of potential and so people are keen to conserve this environment.'

In 2010, Almaden discovered Ixtaca, a gold and silver deposit in the municipality of Ixtacamaxtitlán. Heralded by the company as 'one of Mexico's premier precious metal discoveries,' if developed, Ixtaca will be the first modern metal mine in the state of Puebla (Almaden Minerals, n.d., a; Ehrenberg, 2020). By the end of its fourteen-and-a-half-year lifespan, the open pit will reach over 133 hectares in size – just a shade smaller than Hyde Park – from which Almaden will have extracted 73 million tonnes of mineral (Corporación Ambiental de México, 2018, p. 1).

But Almaden has struggled to obtain the so-called 'social licence to operate', in a region of Mexico where grassroots organization is very strong. Just two years after Almaden announced the discovery of Ixtaca, community resistance in the nearby towns of Tetela de Ocampo and Zautla blocked development of two major mining projects: a gold concession held by Frisco, a company belonging to the billionaire Carlos Slim (then the world's richest man), and a gold, silver, copper, and nickel concession owned by Chinese company JDC. In Ixtacamaxtitlán, local groups have also mobilized, supported by NGOs and academics.

Almaden has been determined not to let community resistance sink Ixtaca and so, over the last decade, alongside its exploration work for the mine, the local subsidiary Minera Gorrión has invested significant time and resources in community engagement and CSR initiatives.

According to the United Nations, CSR is 'a management concept whereby companies integrate social and environmental concerns in their business operations and interactions with their stakeholders' (UNIDO, n.d.). For its proponents, CSR is about companies being good citizens, contributing to the communities and the societies in which they are active, while also mitigating risks, enhancing their reputation, and ultimately improving results (Rangan et al., 2015).

What has Almaden's CSR consisted of in Ixtacamaxtitlán? And to what extent has it been successful?

'Pro Ixtaca'

Alejandro Marreros is from Tecoltemi, a community in the municipality of Ixtacamaxtitlán. He is a member of the Consejo Tiyat Tlali, a network of social organizations in the Sierra Norte de Puebla which has been one of Minera Gorrión's most steadfast opponents.

'In Ixtacamaxtitlán, the company has a whole strategy aimed at co-opting people. And they do it by segments of the population,' he explains. 'In Santa María [Zotoltepec] and Zacatepec, which are the two communities which will be most affected by the open pit if the mine goes ahead, they have programmes aimed at children. They organize summer schools, they take them to theme parks, they buy them school equipment, they organize reading groups and so on. For the women, there are sewing workshops and Zumba classes. They've donated wheelchairs to the elderly.'

Healthcare has been a central focus of Minera Gorrión's CSR work. According to its website, the company donated five vital signs monitors to the hospital in Ixtacamaxtitlán and an ultrasound machine to the clinic in Santa María Zotoltepec, as well as carrying out an eye health campaign amongst local children. It has also used the Covid-19 pandemic as an opportunity to extend CSR initiatives, distributing hand sanitizer and masks, and providing an oxygen tank to the clinic at Santa María Zotoltepec.

'It's a way of whitewashing their image, given that the company was most to blame for bringing Covid into the community,' says Marreros, referring to an outbreak amongst Minera Gorrión's workers in early September 2021, which forced the company to suspend work for two weeks. 'On their social media they said they were a socially responsible company and were suspending operations to protect the health of their workers,' he adds. 'But they only did it because there were already several positive cases, so it was less about prevention and more about control.'

Another initiative of note has consisted of all-expenses-paid trips to visit mining projects elsewhere in Mexico. Minera Gorrión's website states that the company ran 23 such excursions between 2012 and 2017, taking 460 people to learn about 'modern and responsible mining' in five different Mexican states.

'They choose people who are strategic within the community,' says Marreros. 'They put them on a bus and take them to visit the mines and obviously they only show them what they want to show them … They've even been to mines that have terrible track records. They went to one in the state of Oaxaca where there have been killings and things of that sort. But the company tells them that everything's fine, that there's nothing to worry about.'

Those who go on these trips and benefit from other CSR initiatives are then expected to become advocates for Ixtaca within the community.

'Let's say they give you paint for the school,' explains Marreros. 'A couple of weeks or a month later, they'll turn up and say "Look, we've supported you. Now we want you to support us. We don't want anything material, we just want you to come and see what we're doing at the mine site." So they'll take you up there and say "Do you see any pollution? No. Do you see any danger to the community? No. So go and tell people there's no danger, that everything's fine." That's how they reclaim what they donate, and they can be quite demanding about it.'

In this sense, the name Minera Gorrión has chosen for its community engagement strategy in Ixtacamaxtitlán is striking: 'Pro Ixtaca'. In other words, the recipients of Minera Gorrión's CSR benefits become supporters of the mine by default. One photo of a tree-planting programme the company carried out shows a group of children holding a banner which reads 'We are Pro Ixtaca' (Flores, 2018).

Images and videos such as these become PR content, uploaded to the company websites and social networks, featuring in mining industry media, and given visibility at events at which the company is present. Almaden produces this content in English as well as Spanish and it is obviously aimed as much at Almaden's investors as it is at a Mexican audience.

It can be difficult for local people to resist CSR initiatives, particularly when companies promise to satisfy needs which have long been ignored by the state. Still, Minera Gorrión's work in Ixtacamaxtitlán is a good example of how CSR always comes at a cost. Consciously or not, the recipients are drawn into a sophisticated operation designed to give mining projects a veneer of social legitimacy, both within the community and in the outside world.

The employment myth

But what is arguably more persuasive than any CSR initiative is the promise of employment. This is true not only for young men, who are the most likely candidates for any available jobs, but also for their family members and others in the community hoping to benefit indirectly: restaurant owners, hoteliers, shopkeepers and so on. In rural areas like Ixtacamaxitlán, where career opportunities are limited, mining opponents are likely to find this argument particularly difficult to counter.

'They [Minera Gorrión] also managed to turn the heads of some in the community by spinning the line that the company was creating jobs,' says Fernanda Hopenhaym of PODER. 'They take advantage of the fact that these are poor communities … dedicated to cultivating certain products, small-scale trade, and so on … Obviously there are certain economic needs which the company promises to satisfy.'

As of April 2022, Almaden claims on its website to have employed 70 people from local communities. If the mine is given the go-ahead by Mexican authorities, 600 jobs are promised during construction of the mine, falling to 420 by the fifth year of operations (Corporación Ambiental de México, 2018, p. 1).

But the reality is that mining today provides very little in the way of employment. For reasons of both safety and efficiency, technology has replaced much of the workforce in the sector. Most of the employment is short term, available only during the construction phase. Once mines are up and running, demand for labour is minimal, and is largely for skilled professionals who are hired not from local communities, but from elsewhere in the country and sometimes from abroad (Bebbington et al., 2013, pp. 254–255).

According to official figures, 367,935 people were employed in mining in Mexico in 2021 (Gobierno de México, 2022), which is around just 0.66 per cent of the economically active population (Márquez, 2021). Even in Peru – where the sector is more dominant than almost anywhere else in Latin America – it provided just 227,407 direct jobs in the same year (Ministerio de Energía y Minas, 2022), accounting for less than 1.4 per cent of the working population (Instituto Nacional de Estadística e Informática, 2021). Yet the mining sector receives 15 per cent of all foreign direct investment (FDI) in Mexico (Cullell, 2022a) and 25 per cent in Peru (Cruz, 2021), – a clear illustration of the capital-intensive nature of the industry today (PODER et al., 2017, p. 68).

In rural communities like Ixtacamaxtitlán, mining companies usually need to provide employment for several hundred unskilled workers at least; otherwise, they risk them becoming opponents of the mine who might prevent them from ever getting a drill in the ground. But the problem is that there isn't enough work to go around. What companies usually do is to allocate jobs on a rotating basis to distribute the work as widely as possible. Indeed, this is precisely what has happened in Ixtacamaxtitlán.

'They're not permanent jobs,' says Alejandro Marreros. 'They're staggered in order to give a chance to lots of people. One group of workers will be employed for a few days, then another group.'

Companies can offer much higher wages than the local average, so despite the precarity of the employment, workers often remain hopeful that their turn will come round again and that perhaps next time they will be kept on for longer.

'This is an important point, because here the wages are for day labourers, people who work on the land,' adds Marreros. 'It's badly paid. So if a labourer is earning five dollars a day and then a mining company turns up and offers him 25 dollars, then obviously the extra cash is going to be welcome.'

It is a strategy designed to buy the company time, particularly during the crucial phases of exploration and construction, when mining projects are most vulnerable to community resistance. In the medium to long term, these sorts of manoeuvres tend to generate discontent, as people realize the company never had any intention of providing the steady employment it had promised. But by that point it is usually too late.

'They're occupying all the social spaces'

Minera Gorrión's community engagement goes further still. For Marreros, the company has come to dominate what were once traditional social spaces and community occasions in Ixtacamaxtitlán, such as the Day of the Dead celebrations.

'This is a very, very important event for us Mexicans. In fact, when it comes to our festivities and celebrations, it's the most important. It's a sacred time, a sacred space,' says Marreros. 'Now, who is it that organizes the *concurso de las calaveras*?[2] It's the company. Who is it that organizes the *concurso de las ofrendas*?[3] It's the company.'

'At Christmas time, the company is there, sponsoring Las Posadas.[4] On the day of the Epiphany, they're there giving out toys to children. They're there on St. Valentine's Day ... On Children's Day ... On Mothers' Day, Fathers' Day. They're occupying all the social spaces of the community, effectively appropriating them.'

Minera Gorrión has also come to dominate local conversations via its monthly meetings with communities, which were held from 2014 until the outbreak of the Covid-19 pandemic (they restarted in April 2022). Several videos on the company website show footage from these meetings; the person leading almost all of them is Daniel Santamaría, the company vice-president.

'In theory, they're spaces they created to inform the population. Or rather, to dialogue with the population, that's the word they use,' says Marreros. 'But really it's a monologue. For most of the meetings the one who does the talking is Daniel. And what gets talked about is the mining project; it's not possible to discuss anything else.'

What Almaden and Minera Gorrión frame very pointedly in terms of the human right to information, dialogue, and transparency, is really about persuasion. The companies refuse to admit the possibility that the communities don't want Ixtaca, and so the objective is to convince people that mining is in their best interests, gradually chipping away at the culture of these rural and agricultural communities until they become *pueblos mineros* – mining towns.

'There's no community agenda. There's no conversation about what matters to the community,' Marreros adds. 'With these monthly meetings, what the company is saying is that the community agenda now revolves around the mining project. It's the appropriation of the community's social space, the space where people can talk about their problems.'

Despite these regular meetings, PODER's research in Ixtacamaxtitlán suggests the information has not been accurately conveyed in a form which locals can digest. A report on Ixtaca's impact on human rights published in 2017 stated that most locals still did not understand what the project would involve or how they might be affected (PODER et al., 2017, p. 71).

The most egregious example of corporate capture of public space came in June 2019, when the Ministry of the Environment (SEMARNAT) held a meeting in Santa María Zotoltepec to discuss Minera Gorrión's Environmental Impact Assessment (EIA), following requests by the communities and their allies. This was an important meeting, the first time Minera Gorrión's representatives had been obliged to address Ixtaca's opponents directly (Ayala Martínez, 2019). It was also a rare opportunity for communities to raise their concerns with SEMARNAT, the authority responsible for analysing and approving – or not – Ixtaca's EIA. The communities presented 17 objections to the EIA, covering various social, environmental, and technical aspects of the project.

But in a blatant attempt to marginalize Ixtaca's critics and create a false impression of popular support for the project, Minera Gorrión bussed in hundreds of people from outside Ixtacamaxtitlán to attend the meeting. They were dressed in canary yellow T-shirts and baseball caps with slogans reading 'Yes, we want the mine and work in Ixtacamaxtitlán', 'Yes to the Ixtaca project' and 'Yes to work in Ixtacamaxtitlán' (ibid.).

Ixtaca's opponents also complained that Minera Gorrión's statements during the meeting were vague and evasive, focusing on the supposed benefits of mining and the company's CSR efforts in the communities, rather than on what the meeting had been called to discuss: the likely environmental impacts of the mine.

'We asked for a space in which to participate and what we got was company propaganda, which is what we've had for years,' Marreros told PODER following the event. 'We heard the same ready-made phrases, information without any basis. We might as well just have watched one of their corporate videos' (PODER, 2019).

Erasing the indigenous

Among the issues that the communities raised during the meeting was that of consultation with indigenous peoples living within Ixtaca's area of influence. This is often a delicate issue for companies, due to a UN convention called the Indigenous and Tribal Peoples Convention of the International Labour Organization (ILO), better known as ILO Convention 169.

Mexico was one of the first countries to ratify the Convention, in 1990. Since then, it has been legally binding, meaning that indigenous communities must be consulted by the state before any work that might affect their territory can be carried out. Though it falls short of giving indigenous communities the right to veto projects outright, purely formal consultations in which they have no influence over decisions are not sufficient and may be considered a breach of international law (ILO, 2013, p. 16). And while ILO 169 is binding only on states – not companies – private sector actors which are seen to benefit from breaches of the convention may still face legal difficulties, not to mention considerable damage to their reputations (ibid, pp. 25–26).

In Ixtacamaxtitlán, Minera Gorrión has attempted to dodge the issue by simply denying the existence of indigenous communities in the area altogether. In 2017, the company commissioned the firm GMI Consulting to produce a 'social impact assessment'. Required for energy projects in Mexico, but not for mines, these reports are supposed to map the population within a given project's area of influence, assess potential impacts, and help to develop mitigation measures (Almaden Minerals,. n.d.,b). In this case, there was an obvious conflict of interest: the report was financed by the company and produced by a consulting firm which caters to the mining, energy, and infrastructure sectors. It has never been published or made available to the communities.

What *was* made public, was the conclusion that there were no indigenous people living within Ixtaca's area of influence – contradicting existing studies by several government agencies (Flores, 2018).

Then in 2019, in response to a request from Marreros's community of Tecoltemi, which identifies as an indigenous Nahua community, the National Institute of Indigenous Peoples – a state institution – published a statement affirming the existence of 71 localities with indigenous presence in Ixtacamaxtitlán. Nine were located within Ixtaca's direct area of influence (Fundar, 2019).

Tecoltemi has led the fight to be consulted on the project. In 2015, the community filed an injunction against the Ministry of Economy for having granted Minera Gorrión two concessions for Ixtaca which overlapped with its territory, alleging a violation of its rights as an indigenous community (Fundar, 2020). Minera Gorrión argued that Tecoltemi was not indigenous and its claim to be so was 'malicious' (Flores, 2018).

In 2019, a court in Puebla ruled in favour of the community and the concessions were suspended. What Minera Gorrión did in response is revealing. In March 2020, three leaders from Tecoltemi attended a meeting in Apizaco, in the neighbouring state of Tlaxcala, on the invitation of Gilberto Gutiérrez, a contact who had visited the community the previous week. The leaders believed the purpose of the meeting was to discuss cooperative projects in Tecoltemi. But when they arrived at the restaurant where the meeting was to be held, they were met by Daniel Santamaría and Rusbelt Vásquez, vice-president and administrator of Minera Gorrión, respectively.

Santamaría and Vásquez offered them 20 jobs exclusively for Tecoltemi, on the condition that they withdrew the legal injunction. When the leaders refused, the Minera Gorrión men raised the offer to between 30 and 35 jobs for the duration of extraction, plus support for a cooperative project in Tecoltemi which Gutiérrez had proposed (Fundar, 2020).

'This is a very clear, very concrete example of how they aim to break the social fabric by trying to buy out and co-opt community leaders,' says Marreros.

Tecoltemi's legal complaint went beyond the specific concessions granted on the community's land. They argued that Mexico's mining law is illegitimate, violating the country's obligations to its indigenous peoples as set out by both ILO Convention 169 and the national Constitution.

Passed in 1992, in parallel with negotiations on the North American Free Trade Agreement (NAFTA),[5] Mexico's mining law gives the industry priority over almost any other use of land (see Article 6 of the legislation). In theory, mining concessions may be granted in national parks, built-up urban areas,

Photo 3.1 Tecoltemi residents study a map of the concessions / © Mónica González Islas/ El País 2022

and on land belonging to indigenous communities and *ejidos* (Cámara de Diputados, 2022). The only exceptions are areas of hydrocarbon operations and electricity transmission and distribution.

Tecoltemi's case was referred to the Supreme Court, which, in a historic decision in February 2022, cancelled the concessions overlapping with the community's land. This is the first time the Supreme Court has cancelled mining concessions for failure to consult indigenous communities, and it could open the door to similar cases elsewhere in the country. However, the Court upheld the constitutionality of the mining law, meaning that the extraordinary privileges the mining industry enjoys in Mexico remain in place – for now (Cullell, 2022b).

All over?

Almaden has played down the significance of the ruling, claiming it has no interest in mining on indigenous land and suggesting it will wait for the Mexican state to conclude its consultation with the affected indigenous communities. But really this latest decision was another major blow to a project which was already on the ropes.

In December 2020, Almaden revealed that its Environmental Impact Assessment for Ixtaca had been rejected by SEMARNAT. The reason Almaden gave at the time was that the EIA did not contain sufficient information about the potential impacts of the project. Later, in a presentation to the 2021

Mexico Mining Forum, Santamaría admitted that the injunction brought by Tecoltemi had also been a factor (Mexico Business, 2021).

'I think that this is really an unprecedented victory,' says Fernanda Hopenhaym. 'In general, EIAs just don't get rejected here. It was all due to the admirable efforts of the communities in question.'

But Minera Gorrión remains active in the region. Communities in Ixtacamaxtitlán have complained that despite the rejection of the EIA, the company has been doing new exploration work without the requisite environmental licenses. They have asked the authorities to investigate (PODER, 2021).

Over the years, Almaden has consistently downplayed the difficulties it has faced in Ixtacamaxtitlán in its public statements, failing to disclose the significant level of community opposition to the project, or, when that became unsustainable, blaming meddling by 'foreign-funded anti-development NGOs' (Almaden Minerals, 2020). True to form, Almaden signalled that it intends to press ahead with the project despite the verdict from SEMARNAT. As of October 2021, it was preparing a new EIA to address the criticisms raised by the authorities. But Hopenhaym is sceptical.

'The company would have to produce another EIA from scratch, and I don't think they will. We have made this clear to Almaden's investors and shareholders, because when the company talks to the stock exchange it says it's going to appeal the decision,' she says. 'But it's not like they'll be able just to make a cosmetic improvement to the EIA. They would have to start from zero, with much better mitigation measures, and much clearer and more precise information. So I honestly think it's all over.'

II. San Juan, Argentina

The bus to Jáchal from the city of San Juan takes a little over two hours. Sheep graze in fields fringed with bushes and shrubs; the green contrasts with the stark grey forms of the Andes rising up in the distance. The scenery gets rockier and more mountainous the further north I travel.

A sleepy town of just over 20,000 inhabitants, Jáchal is at the centre of what has historically been an agricultural region. It's best known for onion production, though local farmers also produce olives, tomatoes, quince and alfalfa. The Plaza San Martín – the town's central square – is smart, with manicured lawns lined by tall palms. But what no visitor to Jáchal can fail to notice is the large blue tent at one corner of the square, directly facing the town hall across the street. Above the doorway hangs an Argentine flag sporting the words 'ASAMBLEA JÁCHAL NO SE TOCA', which can be translated roughly as the 'Hands Off Jáchal Assembly'.

On one side of the doorway there is a photo of Pope Francis holding up a T-shirt reading 'Water is worth more than gold'; on the other is a kind of rogues' gallery entitled 'Jachal's executioners', featuring among others, Alberto Fernández, Argentina's current president; his two predecessors, Mauricio Macri and Cristina Fernández de Kirchner; the current and former governors of the province of San Juan, Sergio Uñac and José Luis Gioja; and Mark Bristow, the president and CEO of Canadian mining giant Barrick Gold.

Ducking inside the tent, stacks of stools and folded deckchairs line walls decorated with dozens of photos, mostly of members of the Assembly at events and demonstrations. There are placards and banners emblazoned with slogans like 'Barrick Out!', 'Protest isn't a crime; polluting is' and 'We don't want to be a ghost town.' There is also a large cooking pot on a pedestal, the town's tribute to the *cacerolazos* [pot-banging protests] during Argentina's economic crisis of the early 2000s. This was a period of unprecedented popular disgust with the political establishment across Argentina, characterized by the slogan *¡Que se vayan todos!* or 'Down with them all!'

For 58 nights in early 2002, residents of Jáchal took to the streets with their cooking pots to demand the removal of the mayor and his allies on the local council, who had embezzled public funds and left the town in debt. The protests were successful: in late February of that year, the provincial government intervened, dissolving the local council (Zeghaib, 2012). The pot is there to serve as a reminder; a plaque on the pedestal reads 'Dear councillors of Jáchal, the pot is watching you.' Though the pot predates the tent, it has strong symbolic value and it is no coincidence that the Assembly chose the same spot to set up camp.

The atmosphere here is easy-going, peaceful. People sit on deckchairs outside the tent, drinking *mate* – the bitter, caffeine-rich herbal tea ubiquitous in Argentina – and passing round plastic plates stacked with cakes and crackers. Passers-by stop to say hello.

Photo 3.2 Second People's Water Summit in Jáchal / © Oscar Martinez 2019

But since those days of protest in the early 2000s, Jáchal, like other communities in the north of San Juan, is a town transformed.

The 'Second Reconstruction' of San Juan

Unlike neighbouring Chile, Argentina had no history of large-scale industrial mining until relatively recently. Bajo de la Alumbrera, Argentina's first open-pit mine, didn't begin production until 1997, up in the mountainous north-western province of Catamarca. But during the 1990s, a flurry of reforms by the government of Carlos Menem (1989–99) sought to open up Argentina's considerable and largely untapped mineral resources to foreign mining companies. José Luis Gioja, then a deputy for the province of San Juan, was one of the main architects of these reforms, having been appointed by Menem to lead the Congressional Mining Committee.

Gioja became governor of San Juan in 2003 and quickly set about putting Argentina's new mining laws into practice. Harking back to the province's reconstruction following a major earthquake in 1944, Gioja called his project 'the Second Reconstruction of San Juan'. It involved a complete overhaul of the provincial economy, replacing a declining agricultural model based largely on wine production – the dominant industry in San Juan for around a century – with one based on large-scale mining, an activity entirely without precedent.

'The narrative was that this was about bringing San Juan into the 21st century, about transforming a remote mountain area, historically displaced from the Argentine economy, into the number one producer of gold and silver in the region,' says Carina Jofré, an indigenous Warpe activist, assistant researcher at the National Scientific and Technical Research Council (CONICET), and professor at the Universidad Nacional de La Rioja.

In little over a decade, San Juan became one of Argentina's two principal mining provinces, along with Santa Cruz in the far south. Barrick Gold arrived in the early 2000s, aiming to install two major projects high up in the Andes: the gold and silver mine Veladero, and an enormous gold–silver–copper project called Pascua Lama, with operations on both the Chilean and Argentine sides of the *cordillera*. Both projects are located within the San Guillermo Biosphere Reserve, an area created by UNESCO to protect local biodiversity, particularly vicuñas and guanacos, the two wild South American camelids.

Pascua Lama never made it to production. Plagued by social, environmental and technical difficulties from the start, a Chilean court eventually ordered its 'total and definitive closure' in 2020. Veladero, however, came online in 2005 and is projected to run until at least 2030.

The project is located in Iglesia, the department adjacent to Jáchal. Both occupy a remote, peripheral region of San Juan, which had suffered a long period of economic stagnation prior to the arrival of mining. Local agriculture had been in decline throughout the 1980s and 1990s and most young people tend to migrate in search of job opportunities.

'We had one agro-industrial factory, which made sweets, tomatoes in jars and other conserves. But that shut down in 1995,' says Saúl Zeballos, a chartered accountant and founding member of the Asamblea Jáchal No Se Toca. 'In 1996 they declared Jáchal a *zona franca*, a free trade zone, to encourage industries to set up here. But it never happened, the businesses never came. Then in 1997 they promised to double the cultivable area in Jáchal, from 9,000 to 18,000 hectares ... But they never built the network of canals that would have made it possible.'

'This meant that when Barrick arrived, the only thing that anyone wanted was to hand them their CV, to get on their books and to start working.'

Dependent relationships

Like in Ixtacamaxtitlán, Barrick's presence in the north of San Juan has not brought steady employment, though Veladero has long been operational. Instead, the industry has generated a great deal of menial, temporary employment with local municipalities, paid for out of mining royalties. The Asamblea Jáchal No Se Toca calls them the *contratos basura* – literally, the 'rubbish contracts'.

'When they [Barrick] arrived, in 2002 or 2003, they promised zero unemployment. Instead, what we've got in Jáchal are about 1,000 people on contracts with the municipality,' says Zeballos. 'Why are they rubbish? Because they don't pay enough to cover even basic food expenses. And because they're rolling contracts which renew every three months.'

In Iglesia, the situation is even worse than in Jáchal.

'It's well known here that in Iglesia, 90 per cent of the population is on some kind of contract with the municipality. This is brutal, evidence of a labour market which is enormously depressed,' says Jofré.

This is a good example of how mining royalties – though insignificant in comparison with the profits mines like Veladero generate for companies and shareholders – can still distort the public finances of municipalities located close to mines, often rural towns unused to managing large budgets. Not only is it often difficult for municipal authorities to spend this money well, it can also poison local politics, with different groups and actors fighting for control of these resources (Bebbington et al., 2013, pp. 257–258). And of course, there is a risk of corruption.

In the cases of Jáchal and Iglesia, Barrick's presence in the region has created a culture of dependence amongst the local population. Far from the 'sustainable development' trumpeted by Barrick, there are major concerns about what will happen to these towns when the mine closes and the royalties dry up.

'This is a place where traditionally people were self-sufficient. They had their plot of land, their animals,' says Jofré. 'But now, the men – especially the young – go out to work, so it's the women who have to take care of things. People's plots are becoming increasingly small, because the family no longer has the capacity to reproduce this small economy. After a few years, the entire family ends up depending on a grant or a wage from the municipality that comes from the mining royalties.'

Employment has also been used as a means of silencing criticism of mining, as anyone on municipal payrolls who speaks out against the industry or the state runs the risk of being laid off and blacklisted. For example, in 2017 and 2018, at the behest of the mining lobby, Mauricio Macri's centre-right government attempted to relax Argentina's glaciers law, which prohibits activities which might damage glacial and periglacial environments. In Jáchal, any municipal employees who opposed the move had to be discreet about it – or risk losing their job.

'In 2018 we were gathering signatures for a petition we sent to the president against modifying the glaciers law, and the people who were working for the municipality had to sneak behind the tent to sign it,' says Zeballos. 'Why? Because they didn't want to be seen from the town hall. Miguel Vega, the mayor, had told them that if they were seen to be involved in any activity related to the tent or the Asamblea Jáchal No Se Toca, then they would be laid off. So even the little that they had could be taken from them.'

Take it or leave it

Barrick's CSR initiatives have also been used to silence opposition. Like Minera Gorrión in Ixtacamaxtitlán, Barrick has put funding into a wide range of initiatives: from computer training for local teachers, to improvements in local drinking water infrastructure; from oral health programmes in schools, to tree-planting campaigns.

'They've offered shirts and boots to local football clubs. Building materials for dressing rooms, toilets, a clubhouse,' says Zeballos. 'Working with the Ministry of Health, they created a mobile clinic which visits communities out in the province, providing gynaecological screening services. They look for potential cases of cancer and so on.'

These projects generally come in the form of a public–private partnership: the state provides the relevant permissions and any professionals required; the company provides the funding, materials, and sometimes technical assistance (Godfrid, 2018, pp. 215-216).

Just as companies carefully map the physical geography of a given area to determine where mineral deposits are located, how easily they may be extracted and what the potential impacts might be, they also conduct a similar mapping of the local population. These socioeconomic surveys are conducted by specialists in social sciences and the humanities (ibid, pp. 210–211). Thanks to this research, companies can intervene at the points of greatest need. For example, Barrick has invested heavily in local agriculture in Jáchal and the surrounding area, with farmers having pleaded for years for greater state intervention to help arrest the decline of the sector. With Barrick's intervention, farmers found state authorities far more receptive to their proposals (ibid, p. 216).

It's no surprise that CSR initiatives of this type are often eagerly received by communities. But this doesn't mean people are under any illusions. Many people in mining-affected communities are receptive to CSR not because they believe companies have good intentions, but out of a sense of resignation, a feeling that their projects will go ahead come what may and they might as well extract whatever benefit they can from the situation.

This indicates an asymmetrical power relationship which is really the defining characteristic of CSR and of company–community relations in general. While companies attempt to mask this imbalance with talk of 'building partnerships' (Almaden) or 'being a good neighbour' (Barrick), the reality is that the company is *always* at an advantage, particularly – as is the case in San Juan – when the state has invested so much political capital in mining.

'The terms are always dictated by the more powerful party, which is the company,' says Lucio Cuenca, of the Latin American Observatory of Environmental Conflicts (OLCA). 'This distorts any consultation process or social intervention strategy.'

This means that while community groups may present proposals for projects, it is the company which has the final say on whether a project goes ahead, who will be involved and how much support will be provided (ibid, pp. 219–220). Far from being a partnership between equals, communities face a 'take it or leave it' scenario: either they accept what they're offered, or risk being left with nothing but the toxic legacy the mine leaves in its wake.

The spill

In San Juan, this legacy is already a reality. On the 13 of September 2015 there was a spill at Veladero. Over a million litres of a solution containing cyanide and heavy metals leaked into the Rio Potrerillos, subsequently contaminating four other rivers, including the Rio Jáchal (FARN, 2015). It remains Argentina's worst-ever mining disaster.

But residents in Jáchal found out not from the company, nor from the government, but via a WhatsApp message from one of the workers at the mine denouncing the spill and urging locals not to drink the tap water. The message went viral in a matter of minutes and generated terror amongst the local population. There was panic buying of mineral water (Larreta et al., 2017, p. 303). Parents refused to send their children to school (BBC News Mundo, 2015). Whole families even left town (Larreta et al., 2017, p. 304).

Barrick remained silent on the matter for another 24 hours. Then, on 14 September, it released a statement, in which it admitted that there had been a spill, but denied that it had polluted local rivers. 'Contrary to baseless rumours, there has been no contamination of the watercourses in the Rio Jáchal basin,' it said (BBC News Mundo, 2015).

'But there was contamination,' says Saúl Zeballos. 'There was contamination with cyanide and mercury, and this has been proven by information that the company itself provided to the courts in Jáchal.'

Photo 3.3 Second People's Water Summit at the Río Jáchal / © Oscar Martinez 2019

Likewise, Barrick attempted to downplay the scale of the disaster. At first, it claimed that only a small quantity of solution had been spilt – not enough to have an impact on human health, or even to be detectable in the area around the mine. Two days later, it u-turned, owning up to a spill of 224,000 litres. Eventually, the company admitted that more than a million litres had been spilt (FARN, 2015; Editorial RN, n.d.).

Barrick was assisted in its attempts to sweep the incident under the carpet by the provincial government. José Luis Gioja himself stated that there was no possibility of any contamination and accused Barrick's critics of manipulating the information on the spill for political purposes (Lucero, 2015). Likewise, most of the provincial media, particularly the *Diario de Cuyo*, San Juan's leading newspaper, acted as a mouthpiece for Barrick in days after the spill (Leiva, 2015). As the facts became clearer, some local commentators adopted a 'bad apple' approach, singling out Barrick for criticism, while still insisting on the importance of mining for Argentina and taking aim at the industry's 'irrational' critics (Turcumán, 2015).

For the people of Jáchal and San Juan more widely, the spill marked the real beginning of the conflict with Barrick. It was in the aftermath of the spill that the Asamblea Jáchal No Se Toca set up the tent opposite the town hall, while there were demonstrations against the company in the Plaza San Martín and roadblocks were set up along the access route to Veladero (Larreta et al., 2017, p. 304).

'What the spill did was … to cast doubt on the model,' says Carina Jofré. 'It made people start to wonder about this illusion, the idea that mining is the only desirable life goal if you want to live and work in San Juan, the only path to modernity and development. It didn't completely discredit mining or turn everyone against it, that didn't happen. But it planted doubt in people's minds, when before that hadn't been possible.'

History repeating itself

The spill at Veladero in 2015 was not an isolated incident. It was followed almost exactly a year later by another spill, when falling ice damaged a pipe carrying cyanide solution – an event concealed for six days by Barrick and the provincial government (Unidiversidad, 2016). Then the following March there was *another* spill, when a pipe ruptured. While Barrick claimed that the pipe was carrying a gold and silver solution and the leak was contained on site, this is contested by the Asamblea Jáchal No Se Toca, who argued that the solution contained cyanide and mercury as well as the precious metals (*Página 12*, 2017).

There have been some superficial changes in response to the spills. Barrick has been less visible in the media and as a sponsor of cultural events.

'It was striking to see how the government stopped pushing content produced by the companies. They withdrew it because they realized it generated resistance,' says Jofré. 'For example, at the Museum of Fine Arts here [in San Juan] they used to have a sign listing all the mining companies that contributed funding to the museum. They got rid of it. It's not that the companies stopped funding the museum, but they got rid of the sign because it was unpopular. They've had to develop strategies which aren't so visible.'

Barrick sold half of Veladero to the Chinese company Shandong Gold in 2017, entering into a joint venture by the name of Minera Andina del Sol. According to Jofré, Shandong has taken a hands-off approach, with Barrick continuing to manage community relations. The provincial government has also adopted a more critical public stance towards Barrick and has hit the company with some minor penalties: a $9.8 million fine for the 2015 spill, and a second fine of $5.3 million relating to the spills in 2016 and 2017 (Foro Ambiental, 2018). There were also brief suspensions of Veladero's activities following the spills.

But there has been little change of any substance. In fact, the provincial government has been steadfast in its promotion of mining, both to a domestic and international audience. As Argentina emerges slowly from the Covid-19 pandemic, mining is now increasingly promoted as a means of economic reactivation.

The focus is now on Lundin Gold's forthcoming Josemaría project, a vast open-pit copper-gold operation praised by provincial governor Sergio Uñac as a 'landmark in San Juan's mining industry' (Minería y Desarrollo, 2021). Like Veladero, Josemaría is located in the San Guillermo Biosphere Reserve, close to the border with Chile. With an estimated investment of $4.1 billion (as of May 2022), it will be the biggest mining project in Argentina. Following the closure of Bajo de la Alumbrera in 2018, it will also re-establish the country as copper producer at a time when demand is high and set to increase in the coming years (see Chapter 7).

'This is just one of 14 extractive projects in the area. It will be the next one to come online,' says Jofré. 'They made the Environmental Impact Assessment available in February this year [2021], in what was a mendacious – I repeat, mendacious – public consultation.'

Lundin's consultation involved making copies of Josemaría's EIA – a 2,600-page report written in scientific and technical language – available at the provincial Ministry of Mining (Gobierno de San Juan, 2021). What's more, the period of this consultation ran from March to May 2021, coinciding exactly with Argentina's second wave of Covid-19 infection. In late May, with cases soaring, most of the country went back into strict lockdown.

'Anyone from Jáchal who was interested had to travel 160 km to the city of San Juan – even further for people from Iglesia – to get a copy of this report,' says Saúl Zeballos. 'You then had to sign a declaration saying you'd received a copy. That was the only way of registering any objections.'

The retreat of democracy

For Jofré, mining as it has been imposed in San Juan is not simply an economic activity like any other: it is what she calls 'a model of governance'.

'They don't come just to exploit the resources, with the state taking care of everything else. No, it's about co-opting any space that the state has been unable to fill,' she says. 'In countries like Argentina, in places that are socially and economically depressed, like the towns up in the *cordillera*, the state has been absent for a long time. It fails to provide even the basics in terms of health, education and so on. So the companies come and they occupy the role of the nation state, the *pater familias*. And they govern.'

To impose this model of governance, Barrick and other companies in San Juan have benefited from a complicit provincial government. Under first José Luis Gioja and then Sergio Uñac, the government of San Juan has ceded space to the industry, supported company CSR programmes, and promoted mining aggressively at public events and in the media. Most of the province's main media outlets have reinforced this model, providing extensive advertising space to companies; censoring negative news on mining, or reporting it only when it is already widely known; and demonizing those critical of the industry.

What this model implies, therefore, is not only an economic reconfiguration in favour of mining, but the retreat of democracy itself.

'Is it really possible to have autonomy, to imagine the autonomy to make your own decisions in the context of all of this? Are we really free to think and do what we want?' asks Jofré. 'When you send your child to school and the school is financed by the mining industry, when their uniform and their school rucksack have the logo of Barrick Gold or Lundin Gold – it's almost impossible.'

***

Quite aside from the pseudo-philanthropic discourse companies use when talking about CSR – the talk of making 'a positive social difference' (Almaden Minerals, 2019, p. 2) or 'sharing the benefits' (Barrick Gold, 2021, p. 40) – CSR and community engagement are today an essential part of laying the groundwork for mining operations. This is something the companies freely admit. As Barrick puts it, 'Building and maintaining a social licence to operate is critical to our success as a business and our long-term sustainability' (ibid, p. 33)

From a company's point of view, a good community engagement strategy is a way of pre-empting social conflict. This means a safer, more peaceful environment in which to work; less chance of reputational damage to the company; and greater certainty for investors. As such, it may devote considerable time and resources to community relations, as we have seen in the cases of both companies featured in this chapter.

'They have to make a big effort to legitimize their activity, as their activity can't legitimize itself,' says César Padilla, of the Latin American Observatory of Mining Conflicts (OCMAL).

That said, it's important to stress that while the resources that companies invest in CSR and community engagement might look like a lot to the recipients, they are typically a tiny fraction of the profits produced by large-scale mining operations. Veladero alone generates annual earnings in the hundreds of millions of dollars (Mining Data Solutions, 2022).

CSR also has a political dimension. Since the 1990s, as mining has expanded and conflicts with communities have proliferated, there have been increasing calls for tougher regulation and enforcement. CSR allows mining companies to claim that the industry is socially and environmentally responsible and that therefore such reforms are unnecessary.

Critics argue it is a case of the fox guarding the henhouse (Kuyek, 2019, p. 163), in that there is no state or other third-party enforcement of CSR standards, there are no penalties for non-compliance, and there are no real grievance mechanisms in place if communities feel that they have been mistreated. There is also no serious assessment of whether CSR interventions even work – that is, whether they provide solutions for the problems they are ostensibly designed to address.

Even when CSR interventions are well planned and implemented, CSR is frequently divisive. It converts its recipients into ambassadors for the company and isolates any opponents, pitting one side against the other and causing bitter disputes within communities, sometimes even within the same family. This demonstrates how CSR and other community engagement strategies are about far more than simply buying people's silence. They are sophisticated acts of social engineering designed to legitimize the company's presence in a given area, neutralize any potential opposition and break existing community bonds – even to transform the community's very sense of identity (Godfrid, 2018, p. 204).

This social engineering has largely been permitted by Latin American states. With limited political will to invest in peripheral, rural, and indigenous communities, governments have been willing to let companies intervene, providing services which ordinarily would be the responsibility of the state. Such interventions go beyond just public relations; rather, they are an expression of corporate power.

In Ixtacamaxtitlán this has been made abundantly clear.

'At the moment, the company is using the building that used to host the town council as their office. And that's very symbolic,' says Alejandro Marreros. 'What they're saying, is "We're the ones in charge now. The one who calls the shots in this community is the company Minera Gorrión."'

It's not only in the provision of social services that companies have come to adopt the functions of the state. The approach is often one of carrot and stick, with companies reserving measures reminiscent of state repression for mining opponents – another clear sign that the advance of large-scale mining projects implies a corresponding retreat of democracy and civil rights.

In Ixtacamaxtitlán, community members have complained of some low-intensity threats and harassment (PODER, 2017). In San Juan the situation is worse, given the aggressive pro-mining stance of the provincial government. Mining opponents have reported surveillance and harassment, those working in the public sector have been threatened with the sack, and media content critical of mining has been censored in the newspapers and on television. Protest has also been criminalized: in 2016, 32 activists were arrested and detained for blocking the access road to Veladero (la tinta, 2016).

Indeed, the experience of communities in San Juan shows that when companies do decide to resort to repression, the state is often more than happy to do the dirty work on their behalf – as we will see in the next chapter.

Notes

1. The *milpa* is a traditional Mesoamerican agricultural system characterized by polyculture. The main crop is corn, accompanied by others depending on the region, such as squash, chillies, and tomatoes.
2. A Day of the Dead tradition, the *calavera literaria* is a satirical verse written to mock a living individual, often playfully fantasising about their death. The subjects are sometimes politicians or other public figures. The *concurso de las calaveras* is a competition, with prizes awarded for the best poems.
3. Another Day of the Dead tradition, an *ofrenda* is a collection of objects placed on ritual display to commemorate a deceased loved one. They are often highly decorative and elaborate. In the *concurso*, prizes are awarded for the most original and creative arrangements.
4. Las Posadas is a religious festival celebrated in Mexico and elsewhere in Latin America from 16 to 24 of December.
5. NAFTA was replaced by the United States-Mexico-Canada Agreement (USMCA) in 2020, which updates many of the provisions in the former agreement.

References

All references to web-based material were checked and still available in November 2022 unless otherwise stated.

All references are listed, with clickable links for your convenience, on the page for this chapter on the Heart of Our Earth website: <https://lab.org.uk/the-heart-of-our-earth/>.

Almaden Minerals (n.d.,a) 'Ixtaca Gold-Silver Deposit'. [online] Available at: <https://www.almadenminerals.com/project>.

Almaden Minerals (n.d.,b) 'Communication — Almaden Minerals'. [online] Available at: <https://www.almadenminerals.com/communication>.

Almaden Minerals (2019) *Ixtaca Project. Corporate Social Responsibility 2019.* [online] Available at: <https://static1.squarespace.com/static/5ed92e4d58c63e6402d44f65/t/60a2af28329fba4e0e2f2f2d/1621274456346/Almaden_CSR.pdf>.

Almaden Minerals (2020) 'Clarification of the Status of Almaden's Ixtaca Project Mineral Claims'. [online] Available at: <https://static1.squarespace.com/static/5ed92e4d58c63e6402d44f65/t/5ee7c6f845a1d73fa2754c9a/1592248059690/AAU-02-27-2020.pdf>.

Ayala Martínez, A. (2019) 'Acarreo e irregularidades durante reunión pública de Proyecto Minero Ixtaca'. [online] Lado B. Available at: <https://www.ladobe.com.mx/2019/06/acarreo-e-irregularidades-durante-reunion-publica-de-proyecto-minero-ixtaca/>.

Barrick Gold (2021) *Barrick Sustainability Report 2021. Investing in a better future.* [online] Available at: <https://s25.q4cdn.com/322814910/files/doc_downloads/sustainability/Barrick_Sustainability_Report_2021.pdf>.

BBC News Mundo (2015) 'Argentina: temor y protestas en San Juan por derrame de cianuro en una mina'. [online] Available at: <https://www.bbc.com/mundo/noticias/2015/09/150914_derrame_cianuro_mina_sanjuan_argentina_bm>.

Bebbington, A., Humphreys Bebbington, D., Hinojosa, L., Burneo, M. and Bury, J. (2013) 'Anatomies of Conflict: Social Mobilisation and New Political Ecologies of the Andes', in A. Bebbington and J. Bury, eds., *Subterranean Struggles. New Dynamics of Mining, Oil, and Gas in Latin America*. Austin, TX: University of Texas Press, pp. 241–266.

Cámara de Diputados del H. Congreso de la Unión (2022) *Ley Minera.* [online] Available at: <https://www.diputados.gob.mx/LeyesBiblio/pdf_mov/Ley_Minera.pdf>.

Corporación Ambiental de México (2018) *Resumen Ejecutivo MIA-Reg con Análisis de Riesgo – Proyecto Minero Ixtaca.* [online] Available at: <https://apps1.semarnat.gob.mx:8443/dgiraDocs/documentos/pue/resumenes/2019/21PU2019M0006.pdf>.

Cruz, E. (2021) 'El 25% de la inversión extranjera directa en Perú proviene del sector minero'. [online] *Rumbo Minero.* Available at: <https://www.rumbominero.com/peru/noticias/mineria/el-25-por-ciento-de-la-inversion-extranjera-directa-en-peru-proviene-del-sector-minero/>.

Cullell, J.M. (2022a) 'La inversión extranjera en México aumenta un 9% en 2021 pero se mantiene en niveles de hace un lustro'. [online] *El País.* Available at:

<https://elpais.com/mexico/economia/2022-02-21/la-inversion-extranjera-en-mexico-aumenta-un-9-en-2021-pero-se-mantiene-en-niveles-de-hace-un-lustro.html>.

Cullell, J.M. (2022b) 'La Suprema Corte cancela dos concesiones mineras en Tecoltemi en una decisión histórica'. [online] *El País*. Available at: <https://elpais.com/mexico/2022-02-16/la-suprema-corte-cancela-dos-concesiones-mineras-en-tecoltemi-en-una-decision-inedita.html>.

Editorial RN (n.d.) 'El derrame, Barrick y su comunicado'. [online] Editorial RN. Available at: <http://editorialrn.com.ar/index.php?option=com_content&view=article&id=3366:el-derrame-barrick-y-su-comunicado&catid=14&Itemid=599>.

Ehrenberg, A. (2020) 'Almaden's Ixtaca: Sustainable Economic Development for Puebla'. [online] Mexico Business News. Available at: <https://mexicobusiness.news/mining/news/almadens-ixtaca-sustainable-economic-development-puebla>.

Flores, N. (2018) 'Trasnacional Almaden Minerals "borra" indígenas para evadir consulta'. [online] *Contralínea*. Available at: <https://contralinea.com.mx/trasnacional-almaden-minerals-borra-indigenas-evadir-consulta/>.

Foro Ambiental (2018) 'Barrick Gold recibe una nueva multa por sus derrames'. [online] Available at: <https://www.foroambiental.net/barrick-gold-recibe-una-nueva-multa-derrames/>.

Fundación Ambiente y Recursos Naturales (2015) 'Cronología del derrame en Veladero'. [online] Available at: <https://farn.org.ar/cronologia-del-derrame-en-veladero/>.

Fundar (2019) 'Hay población indígena en el área del Proyecto Minero Ixtaca: INPI'. [online] Available at: <https://fundar.org.mx/inpi-poblacion-indigena-ixtaca/>.

Fundar (2020) 'Minera intenta sobornar a comunidad para que desista de amparo'. [online] Available at: <https://fundar.org.mx/almaden-minerals-intenta-sobornar-a-comunidad-indigena-para-que-se-desista-de-amparo/>.

Gobierno de México (2022) 'Minería'. [online] Available at: <https://www.gob.mx/se/acciones-y-programas/mineria>.

Gobierno de San Juan (2021) 'Lanzaron la apertura a la convocatoria a consulta pública de Josemaría'. [online] Available at: <https://sisanjuan.gob.ar/mineria/2021-03-05/30140-lanzaron-la-apertura-a-la-convocatoria-a-consulta-publica-de-josemaria>.

Godfrid, J. (2018) 'La implementación de iniciativas de responsabilidad social empresaria en el sector minero. Un estudio a partir de los casos Alumbrera y Veladero', in L.A. Huwiler and J. Godfrid, eds., *Megaminería en América Latina: Estados empresas transnacionales y conflictos socioambientales*, pp. 199–228.

Instituto Nacional de Estadística e Informática (2021) *Ocupados a nivel nacional alcanza 16 millones 848 mil 600 personas en el II trimestre de 2021*. [online] Available at: <http://m.inei.gob.pe/media/MenuRecursivo/noticias/nota-de-prensa-no-122-2021-inei.pdf>.

International Labour Organization (2013) *Understanding the Indigenous and Tribal Peoples Convention, 1989 (No. 169). HANDBOOK For ILO Tripartite Constituents*. [ebook] Geneva: International Labour Organization. Available at: <https://www.ilo.org/wcmsp5/groups/public/---ed_norm/---normes/documents/publication/wcms_205225.pdf>.

Kuyek, J. (2019) *Unearthing Justice. How to protect your community from the mining industry*. Toronto: Between The Lines.

la tinta (2016) 'Jáchal: bloqueo y represión'. [online] Available at: <https://latinta.com.ar/2016/12/jachal-bloqueo-y-represion/>.

Larreta, G., Sánchez, V., Donoso Ríos, M. and Rodríguez Ruarte, M. (2017) 'Asamblea Jáchal No Se Toca: Crónica de una catástrofe', *RevIISE*, 10(10), pp. 299-312. Available at: <https://ri.conicet.gov.ar/bitstream/handle/11336/102962/CONICET_Digital_Nro.00d209a5-f9f5-40ff-8f67-84e79eece965_A.pdf?sequence=2&isAllowed=y>.

Leiva, C. (2015) 'El sismo o el frío, posibles causantes del derrame en Veladero'. [online] *Diario de Cuyo*. Available at: <https://www.diariodecuyo.com.ar/economia/El-sismo-o-el-frio-posibles-causantes-del-derrame-en-Veladero-20150915-0056.html>.

Lucero, J. (2015) 'Gioja dijo que 'no hay contaminación del agua' por el derrame de cianuro en una mina de San Juan'. [online] télam. Available at: <https://www.telam.com.ar/notas/201509/120018-cianuro-san-juan-gioja.html>.

Márquez, E. (2021) 'Cuántos son los trabajadores que hay en México 2021, según el INEGI'. [online] Mi Trabajo. Available at: <https://web.archive.org/web/20210714183556/https://www.mitrabajo.news/noticias/Cuantos-son-los-trabajadores-que-hay-en-Mexico-2021-segun-el-INEGI-20210711-0005.html>.

Martínez, M., Evangelista, V., Basurto, F., Mendoza, M. and Cruz-Rivas, A. (2007) 'Flora útil de los cafetales en la Sierra Norte de Puebla, México', *Revista Mexicana de Biodiversidad*, 78, pp. 15–40. Available at: <http://www.ejournal.unam.mx/bio/BIOD78-01/BIO007800103.pdf>.

Mexico Business (2021) *Operator Presentation Almaden Minerals*. [video] Available at: <https://www.youtube.com/watch?v=OZEzjoHvRjo>.

Minería y Desarrollo (2021) 'Uñac en la Legislatura: minería responsable, Josemaría y laboratorio de minerales no metalíferos'. [online] Available at: <https://mineriaydesarrollo.com/2021/04/01/unac-en-la-legislatura-mineria-responsable-josemaria-y-laboratorio-de-minerales-no-metaliferos/>.

Mining Data Solutions (2022) 'Major Mines & Projects | Veladero Mine'. [online] Available at: <https://miningdataonline.com/property/255/Veladero-Mine.aspx>.

Ministerio de Energía y Minas (2022) 'Empleo directo en minería consolidó un resultado histórico en el 2021'. [online] Available at: <https://www.gob.pe/institucion/minem/noticias/581253-empleo-directo-en-mineria-consolido-un-resultado-historico-en-el-2021>.

Página 12 (2017) 'Un nuevo derrame en la mina Veladero'. [online] Available at: <https://www.pagina12.com.ar/28713-un-nuevo-derrame-en-la-mina-veladero>.

PODER (2017) 'Empleados de Almaden Minerals violentan a habitantes de la Sierra Norte de Puebla'. [online] Available at: <https://poderlatam.org/2017/05/empleados-de-minera-almaden-minerals-violentan-y-amenazan-a-habitantes-de-la-sierra-norte-de-puebla/>.

PODER (2019) 'Irregularidades en reunión pública de información en Ixtacamaxtitlán'. [online] Available at: <https://poderlatam.org/2019/06/denuncian-comunidades-de-ixtacamaxtitlan-irregularidades-y-manipulacion-en-la-reunion-publica-de-informacion-organizada-por-la-semarnat/>.

PODER (2021) 'Habitantes de Ixtacamaxtitlán denuncian ante Profepa actividades de exploración no autorizadas de Almaden Minerals'. [online] Available at: <https://poderlatam.org/2021/08/habitantes-de-ixtacamaxtitlan-denuncian-ante-profepa-actividades-de-exploracion-no-autorizadas-de-almaden-minerals/>.

PODER; Unión de Ejidos y Comunidades en Defensa de la Tierra, el Agua y la Vida, Atcolhua; CESDER; IMDEC (2017) *Minería canadiense en Puebla y su impacto en los derechos humanos.* [online] Available at: <https://poderlatam.org/wp-content/uploads/2020/02/InformeMineriaCanadiense-2017_SNP.pdf>.

Poliquin, M.J. (2009) *Geology, Geochemistry and Age of Intrusion-Related Mineralisation in Eastern Mexico* [online] PhD. University of Exeter. Available at: <https://ore.exeter.ac.uk/repository/handle/10036/108354>.

Ramos Mancilla, O. (2015) *Internet y pueblos indígenas de la Sierra Norte de Puebla, México.* [online] PhD. Universitat de Barcelona. Available at: <https://www.tdx.cat/bitstream/handle/10803/353624/ORM_TESIS.pdf?sequence=2&isAllowed=y>.

Rangan, V., Chase, L. and Karim, S. (2015) 'The Truth About CSR'. [online] *Harvard Business Review.* Available at: <https://hbr.org/2015/01/the-truth-about-csr>.

Turcumán, J. (2015) 'El drama de confiar en una mina mentirosa'. [online] *Diario de Cuyo.* Available at: <https://www.diariodecuyo.com.ar/columnasdeopinion/El-drama-de-confiar-en-una-mina-mentirosa-20150920-0097.html>.

Unidiversidad (2016) 'Greenpeace le pegó a Bergman por Veladero'. [online] Available at: <https://www.unidiversidad.com.ar/ante-la-denuncia-de-bergman-greenpeace-volvio-a-exigir-el-cierre-de-veladero>.

United Nations Industrial Development Organization (n.d.) 'What is CSR?' [online] Available at: <https://www.unido.org/our-focus/advancing-economic-competitiveness/competitive-trade-capacities-and-corporate-responsibility/corporate-social-responsibility-market-integration/what-csr>

Zeghaib, M. (2012) 'El monumento sanjuanino al cacerolazo'. [online] *Tiempo de San Juan.* Available at: <https://www.tiempodesanjuan.com/sanjuan/2012/10/7/monumento-sanjuanino-cacerolazo-18833.html>.

CHAPTER 4
Resources before rights

Jo Griffin

Community resistance to mining generates a reaction from companies and governments, often resulting in criminalization and human rights abuses. This chapter presents a recent case of criminalization of mining opponents in Honduras; it then tells the story of the systematic violation of the human rights of Afro-descendant and indigenous Wayúu communities by the coal miner Cerrejón, in La Guajira, Colombia; finally, it examines human rights abuses during the conflict between local communities and the Yanacocha mining company in Cajamarca, in the Peruvian Andes.

I. Guapinol, Honduras: impunity rules

'My father spent almost three years locked up and having him home is an enormous joy for me and my family,' says Gabriela Sorto. 'But at the same time, we know the struggle is not over. The mining company continues to pollute and destroy our river, which is drying up every day, and we will keep fighting for them to leave the area and for our right to live in a clean environment.'

Sorto's father, the farmworker Porfirio Sorto Cedillo, had languished in pre-trial detention with seven others from September 2019 to February 2022, accused of criminal activity during demonstrations against an iron ore mine in Guapinol, a village just outside Tocoa, around 210 km north-east of the capital Tegucigalpa.

The case of the 'Guapinol Eight' was another twist in a decades-long land conflict in the once-fertile Bajo Aguán valley, where local campesinos have struggled bitterly against the expansion of big agribusiness. Thanks to a World Bank-sponsored reform in 1992 allowing collectively held plots of land to be carved up and sold, powerful landowners have bought up most of the land in the valley, largely for industrial production of palm oil (Chavkin, 2015). Many of those who resist have been killed or disappeared – at least 150 between 2010 and 2014 alone (Olson, 2020).

And in recent years, locals have also had mining to contend with. In 2014, a mining concession was granted to Honduran company Inversiones Los Pinares (ILP) following a dubious change to the law (ILAS, 2020). In 2018, the company began building a road to the proposed site. Before long, the pristine water of the Río Guapinol had turned chocolate brown and was thick with mud; children were suffering from diarrhoea and skin rashes.

'Our families were getting sick, we had to buy bottled water to drink,' Sorto recalls.

In August of that year, locals from Guapinol took action. About 140 supporters of the Municipal Committee for the Defence of Common and Public Property set up a camp, blocking the road to the mine site. The following month, there was a standoff at the camp with ILP security guards, at which a protestor was shot and seriously injured. ILP claims a vehicle and other company property was set on fire (University of Virginia School of Law, 2020, p. 13).

Following the shooting, demonstrators managed to detain the company's head of security until the police arrived. Yet the incident was never investigated, though it had taken place before dozens of witnesses. Instead, the Public Prosecutor's Office decided to pursue members of the Guapinol community, based on ILP accusations of trespassing and criminal damage (ibid.).

Two arrest warrants were issued: the first, in October 2018, contained relatively minor accusations against eighteen of the water defenders. However, the second, issued in early 2019, accused thirty-one community members of 'arson, theft, false imprisonment and criminal association.' The latter charge is particularly serious; it meant that their cases would be heard not before a local court, but by a court of national jurisdiction dedicated to high profile criminal cases, with a special focus on organized crime (ILAS, 2022).

The Guapinol Eight are seven activists who presented themselves to the courts in August 2019 to respond to the accusations in the second arrest warrant, plus one other man who was arrested in November 2018 on the basis of the first warrant, then kept in pre-trial detention due to the second. To highlight the flimsiness of the case against them, the seven men arrived for their hearing carrying a coffin; it represented Antonio Martínez Ramos, a community member included in the arrest warrant who had died in 2015, long before the alleged crimes took place (ibid.).

Watching her father being dragged off by police was traumatic, says Sorto. 'For us, it was a kidnapping. They seized them and took them away, handcuffed and in their underwear, to La Tolva [a maximum-security prison] in south-eastern Honduras, a long way from where we live.'

The men spent two months in the notorious jail before a judge granted an appeal to move them nearer home. This was a rare victory, in a case that revealed not only how a dysfunctional legal system fails to punish human rights abusers, but how the courts and other state institutions have been mobilized against land and environmental defenders. In particular, lawyers for the eight men argued that there was no justification for their pre-trial detention, a measure which the Honduran Criminal Code reserves for extreme situations (University of Virginia School of Law, 2020, p. 17).

In February 2021 the UN Working Group for Arbitrary Detention said there was no legal basis to detain the eight men and demanded their immediate release, a call echoed later that year by Honduras' new president-elect Xiomara Castro (OHCHR, 2021, p. 14).

On 9 February 2022, six of the Guapinol Eight were found guilty and sentenced to a total of 110 years in prison (ILAS, 2022). However, the very next day the Supreme Court issued an unexpected ruling accepting appeals filed six months beforehand, which had challenged the constitutionality of the charges and pre-trial detention. Two weeks later, they were freed.

But this victory has come at a price. The Guapinol camp was removed in October 2018 by 1,500 soldiers and police officers – more than double the number of demonstrators who had come together to resist the eviction. The security forces used not only tear gas to break up the camp but also live ammunition, killing one demonstrator and wounding many others (University of Virginia School of Law, 2020, p. 13).

Two military personnel also lost their lives in circumstances which remain unclear; according to the police, they were shot from the forest with AK-47 assault rifles (*El Heraldo*, 2018). One human rights investigator has suggested that the killings were the result of a clash between the military and local ex-associates of the Cachiros drug cartel, with the deaths then being used to smear local opposition to the mine (Olson, 2020).

Two other Guapinol water defenders were subsequently killed, both of whom featured in the second arrest warrant. On 28 August 2019, Roberto Antonio Argueta Tejada was shot dead in the street in Tocoa. Little over a year later, on 13 October 2020, Arnold Joaquín Morazán Erazo was shot at home by two gunmen and died of his injuries. Like the other murders and disappearances of campesinos in the Bajo Aguán valley to early 2020, these crimes remain unsolved.

'Having my father home and sharing our daily lives with him again has made us so happy,' says Sorto. 'But we know that the company is angry because they would have liked all the defenders to remain in prison. My father does not go out because he is still frightened of what they might do.'

Climate of fear

Outside La Tolva prison, east of Tegucigalpa, dozens of women line up in the searing heat. There is no shade in front of the imposing front gate, known as 'Customs One', and children grow restless at their feet.

Since it was built in 2017 to ease the pressure on the overcrowded prison system, La Tolva has regularly hit the headlines for riots and deadly clashes between gangs. The human rights organization COFADEH has described the facility as a 'torture centre' (Defensores en Línea, 2019). The fact that the Guapinol water defenders were imprisoned there is an indication of the scale of repression which has faced mining opponents in Honduras.

'They are using jails built for organized crime for the government's opponents. Since 2018 people defending land and water from contamination linked to mining have been sent to these prisons,' says Karen Spring, a Canadian human rights activist who has lived in Honduras for more than a decade.

Not so long ago, Spring was one of the women queuing to enter La Tolva, after her partner, Edwin Espinal, was jailed with twenty-four others for demonstrating against the outcome of the presidential election on 26 November 2017. The incumbent, Juan Orlando Hernández, claimed victory, despite widespread allegations of fraud and other irregularities. According to the UN, there were human rights abuses before and after the controversial poll; at least 23 people were killed, mostly by security services (OHCHR, 2018). Espinal and Raúl Álvarez, another political prisoner jailed at the same time, were released on bail in August 2019, after 19 months in La Tolva. In September 2021 they were cleared of all charges.

'[At La Tolva] there is access to water for just two hours a day and a lot of conflict over bathroom breaks,' says Spring. 'Inmates hardly get any sunlight … conditions are horrendous.'

According to Spring, pre-trial detention for defending water or land has become commonplace in Honduras, with the judicial system having been corrupted to serve mining interests.

'What generally happens is that the attorney general's office is in a very close relationship with the mining or other private company – or even a state institution. They work together to come up with some sort of charge,' she explains. 'It could be a range of things; with a land struggle, it tends to be usurpation or illegal possession of land – or damaging private property. But they have also started to use other charges relating to organized crime, and this is what they did in Guapinol, for example, where they alleged criminal association and conspiracy.'

Jailing a defender – usually the breadwinner – often leaves their family in dire straits. But families are also crippled by the demands on those who remain at liberty but facing charges. Hondurans must pay for their own legal defence, but if they can't, they have to sign before a judge every week, which might mean hours of travel when they can barely afford food.

'If defenders are not sent to prison, they are slapped with criminal charges, such as illegal possession of land,' says the veteran human rights campaigner Pedro Landa. 'Very poor farmers are criminalized and can't pay for a lawyer and then the public defender makes no effort to defend them or to even give the impression it is a fair system. The lawyers who come from human rights organizations put themselves in extreme danger and are a targeted group. Journalists who expose this are also murdered.'

According to the Honduran College of Journalists, at least 86 journalists were murdered between 2001 and July 2020, with 92 per cent of those killings going unpunished (Human Rights Watch, 2021). Police corruption makes it harder to achieve justice and a National Protection Mechanism created in 2015 to protect human rights defenders is ineffective, says Landa. Meanwhile, laws that should protect communities' human rights are circumvented, unenforced or simply ignored.

These include the requirement for consultation with communities before exploration of indigenous land, under ILO Convention 169 (see Chapter 3). Despite Honduras ratifying this in 1995 and voting in favour of the UN Declaration on the Rights of Indigenous Peoples in 2007, many projects have gone ahead without prior consultation, such as dams on rivers sacred to the Lenca, the country's largest indigenous group, and tourist developments affecting the Garifuna, an Afro-indigenous group which lives mostly along the coast.

Rocío Walkiria, of the rural human rights organization CEHPRODEC, says that many communities were 'consulted *after* decisions had already been made about their land.' A new law on consultation in 2018, which has made it possible for companies to ignore the outcome of any consultation, simply cemented 'how things are always done.'

The 'second colonization'

During a lifetime spent fighting for human rights, Landa has received countless death threats, been beaten up in front of a police station, and spent long periods in hiding. 'I have been fighting for justice since the days when Honduras used to export all our fruit to Asia,' he says.

But for Landa, the incursion of mining companies is 'like a second colonization.'

'During the lockdown the situation for communities has got even worse,' he says, referring to the strict measures brought in by the government in 2020 as part of its Covid-19 response. 'Mining and hydroelectric projects were declared "essential activities" that could not be halted by Covid, while protests were banned, and that allowed projects to advance.'

As well as allowing the government to further shrink civic space, the pandemic has also brought back memories of previous crises 'exploited by the state' to rush through policies to encourage mining.

'What we saw [with Covid] is like 1998, when a mining law was rushed through in the wake of Hurricane Mitch, one of the biggest disasters to hit Honduras. Economically, Mitch set us back forty years,' he says.

Passed just four weeks after the storm, the General Mining Law allowed companies to explore the country freely and evict communities from the land they were granted. It also granted them generous tax exemptions, while relaxing environmental controls. The military were deployed in remote areas for 'humanitarian' purposes, a shady practice that continues today (Holland, 2015). With increasing militarization of ancestral lands targeted for 'development', new military forces have been created to patrol areas home to strategic industries, such as mines, manufacturing, and hydroelectric plants (Hammond et al., 2020, p. 15; PBI Honduras, n.d.).

But in the years following approval of the General Mining Law, there was a major backlash. A campaign to overturn the law was coordinated by civil society organizations, international NGOs, and the Catholic Church, fuelled by popular anger at the disastrous impacts of mining in rural areas throughout the country. A Supreme Court ruling in 2006 struck down some of the worst articles in the law and there was hope it would be scrapped altogether. By May 2009 a new law had been drafted, prohibiting open-pit mining, banning the use of mercury and cyanide, and making community approval a requirement for any mining project (Holland, 2015).

That bill would never make it to Congress, however. In the small hours of 28 June 2009, the then president Manuel Zelaya was awoken by soldiers and bundled onto a plane to Costa Rica in his pyjamas, in what the UN, EU, and Organization of American States all condemned as a military coup. It brought the pro-business National Party to power, which quickly set about pursuing a political project which aimed to hand over Honduras' natural resources to foreign multinationals and transfer vast swathes of land to a tiny elite (Gordon and Webber, 2016, p. 83).

The National Party regime enjoyed the support of powerful foreign allies, particularly the governments of Canada and the United States. In 2013, a new mining law was passed, with extensive support from the Canadian state and Canadian mining executives. The law failed to protect communities' access to water, contained loopholes allowing foreign companies to reduce their tax liabilities, and failed to ensure meaningful community consultation (ibid., p. 69).

Again, the Guapinol conflict illustrates how mining interests were given precedence by the National Party regime. In 2011, the nearby Botaderos Mountain was declared a national park, guaranteeing protection of water sources including the Río Guapinol. But after Inversiones Los Pinares made requests to mine within the park, another bill was passed in December 2013, reducing the protected zone by 223.7 hectares to allow space for the mine. A concession was granted to ILP in early 2014 and that same year the project received its environmental licences. At no point were communities consulted (OHCHR, 2019).

ILP is owned by one of the wealthiest couples in the country: Lenir Pérez, a businessman accused of involvement in bribery, intimidation, and kidnapping (Global Witness, 2017, p. 20); and his wife Ana Facussé, daughter of the late Miguel Facussé, a notorious palm oil tycoon who was one of the main players in the violent land grabbing campaign in Bajo Aguán (Beeton, 2012). U.S. State Department cables and testimony by former Honduran cartel members turned DEA informants also suggest that Facussé senior's properties have been used to transport cocaine (Olson, 2020). Facussé, who was Honduras' wealthiest and most powerful businessman, had close ties to the National Party and was a prominent supporter of the 2009 coup.

Xiomara Castro – Manuel Zelaya's wife – won a decisive election victory in late 2021, bringing to an end a period of over twelve and a half years marked by human rights violations, electoral fraud, rampant corruption, and the capture of the state by global organized crime. Juan Orlando Hernández's younger brother Tony was given a life sentence for smuggling tonnes of cocaine into the United States in 2021, while Juan Orlando himself was arrested in dramatic fashion just weeks after stepping down as president and subsequently extradited to the United States to face drugs, weapons, and money laundering charges. Anne Milgram, the head of the DEA, called him 'a central figure in one of the largest and most violent cocaine trafficking conspiracies in the world' (BBC News, 2022).

In February 2022, just a month after taking office, Castro announced a ban on open-pit mining in Honduras, fulfilling a pledge made during the election campaign. The government promised the 'revision, suspension, and cancellation of environmental licenses, permissions, and concessions', though as of May 2022, it is unclear to what extent the government will be able to shut down existing operations, or whether the ban will apply only to new projects (Dinero HN, 2022). Nonetheless, it is a major victory for mining-affected communities and environmental and human rights defenders throughout the country.

Still, the challenges facing the Castro administration are formidable. Poverty, violence, and organized crime all spiralled under the National Party, with Honduras becoming the largest source of migrants to the United States (Call, 2021). Mining has contributed to this: as conflicts have erupted all over the country, hundreds of thousands of people have fled their homes in search of safety.

'Mining is inextricably linked to the migration crisis because people cannot sustain themselves economically,' says Pedro Landa. 'People feel abandoned, orphaned. Thousands of people are just [living] in the streets.'

Honduras remains one of the world's deadliest places in which to defend the environment. Though its population is less than ten million, hundreds of environmental and land defenders have been murdered or disappeared since the coup in 2009. Women have not been spared, as became clear in 2016 when this wave of violence claimed its most high-profile victim.

Women on the frontline

In March 2016, the Goldman Environmental Prize-winning activist Berta Cáceres was shot dead in her own home. For weeks there had been signs her life was in immediate danger because of her campaigning against the Agua Zarca hydroelectric dam, which, had it been built, would have dissected the Gualcarque River, sacred to the Lenca people. She had told colleagues at COPINH, the environmental and indigenous rights organization she founded, that a black SUV had been following her. Though she was used to threats, this time she was nervous, dismissing reassurances her fame would protect her.

'I take precautions, but in the end, in this country where there is total impunity I am vulnerable. When they want to kill me, they will do it,' she said (Lakhani, 2016).

Cáceres' murder put the spotlight on women at the forefront of the struggle in Honduras, leading eight UN special rapporteurs to co-sign a call for an end to impunity (OHCHR 2016). But women continue to be targeted; an analysis

Photo 4.1 Berta Cáceres takes a moment to remember friends and colleagues killed in the struggle against the Agua Zarca Dam / © Goldman Environmental Prize 2015

from the IMS-Defensoras network in August 2021 showed that at least seven women defenders have been killed since 2016 (Davies, 2021).

'The intimidation of women has a chilling effect', says Karen Spring. 'It generates terror when a woman is targeted – women are not seen as dangerous as men and when a woman is targeted it is still shocking. The threat comes directly into the home.'

Not only is Honduras one of the most dangerous countries for activists, it is also one of the world's deadliest places for women. A woman is killed on average every 23 hours, and the perpetrators – who might be their partner, a relative, local gang members or even the police – are almost never brought to justice (Human Rights Watch, 2021). In such a climate, women who speak out in defence of their territories are at even greater risk. What's more, they face being stigmatized and cut off by the very communities they are fighting for.

'Imagine you are "María" at home with three children whom you have to feed, and you have to defend that source of water for them. But María must leave home to defend the water source and a woman who does not stay at home is a "bad woman",' says Rocío Walkiria. 'The impact on women is very different. Sometimes their relatives have jobs [with the mining company], and they will tell a woman to stop speaking out.'

'For us young women there are definitely risks and we get abuse on the streets,' says Gabriela Sorto. 'But it makes us more determined to stand up and defend what is ours.'

In what was welcomed as a rare success in a country where the justice system fails in 90 per cent of cases, on 6 July 2021 David Castillo, the former manager of Desa, the company behind Agua Zarca, was found guilty of planning and ordering the murder of Berta Cáceres. As of May 2022, he had yet to be sentenced, though the seven assassins he hired were convicted in a separate trial in 2018 and received sentences of between 30 and 50 years.

'This is a victory for our *compañera* Berta and her family but also for all communities in the struggle,' says Sorto. 'However, we know there is much more to do.'

II. La Guajira, Colombia: sacrifice zone

For as long as Luz Ángela Uriana can remember, the air around her home on the Provincial indigenous reservation in La Guajira has been thick with dust from the giant open-pit coalmine of Cerrejón, less than 2 km away.

When her son Moisés Daniel was about six months old, he began to suffer from a fever and dry cough, so she took him to the doctor. Her quest to protect the health of her six children is just part of a wider struggle with other indigenous Wayúu, Afro-descendant and campesino communities to defend their families' right to live healthy lives on their ancestral territory on the Guajira peninsula along the Colombia-Venezuela border.

In 2015 a judge ruled in favour of Moisés, whose breathing problems were blamed on dust caused by the mine. Though Cerrejón denied responsibility, both the company and Colombia's environment agency were ordered to provide medical treatment to Moisés and take steps to reduce pollution around the mine. But seven years later, the Wayúu are still fighting for clean air, pushing for the enforcement of a series of Constitutional Court rulings ordering the mining company and authorities to address pollution and other problems in La Guajira.

Fears that poor health had left people in La Guajira dangerously vulnerable to Covid-19 led Uriana and others to call for mining to stop during the pandemic and urge the UN to intervene.

'My son is seven now and he is a bit better because of the medicine he has received, but I've been really worried about his lung problems during Covid,' she says.

For the region's indigenous people, the pandemic added yet another layer of threat to a longstanding and complex crisis that in 2020 UN special rapporteur David Boyd called one of the most 'disturbing situations' he had ever seen (OHCHR, 2020).

The situation in La Guajira has illustrated how basic human rights are placed in jeopardy when an absentee state leaves marginalized people at the mercy of a mining giant. It also highlights the difficulties communities face when they seek justice through the same state that has ceded control of their land and resources to a corporation.

After almost forty years of mineral extraction in La Guajira, the department remains amongst the poorest in Colombia. Communities in the mine's area of influence live in extreme poverty with little access to basic services. Despite multiple Constitutional Court rulings in their favour, they continue to battle with drought and water shortages, food insecurity, and severe health problems linked to pollution. Alarming levels of child malnutrition have worried experts for years; children under five in La Guajira are nearly six times more likely to die from malnutrition than in the rest of Colombia and the Wayúu are the worst affected (Human Rights Watch, 2020).

Rogelio Ustate Arrogoces is the legal representative of Tabaco, an Afro-Colombian community that was forcibly displaced by the mine in 2001. 'We are suffering a catastrophe,' he said in an interview. According to Ustate, disease and death have followed the fundamental alteration of the landscape, with deforestation worsening droughts and around 60,000 tonnes of explosives every year polluting not just surface water but subterranean water too. 'That's why so many old people and children are dying,' he said.

'Whether from deforestation or the loading or unloading of materials, the atmospheric pollution is damaging the whole region ... The dust causes breast, skin, cervical, oesophageal, throat and lung cancer. It damages DNA structure and causes deformities. That's why there are so many children with special needs,' he said.

The 'health crisis is very complex', says the anthropologist Emma Banks, who has worked with communities in La Guajira for years. Pollution, a lack of health services, and food and water shortages have all contributed to deteriorating public health. Climate change is also a factor: much of La Guajira is a coastal desert, historically an extremely hot and dry region, and in recent years the droughts have been increasingly severe.

Cerrejón has long refuted accusations the mine is responsible for the crisis, but in 2021 the Organisation for Economic Co-operation and Development (OECD) increased global pressure on the company by launching an investigation into alleged human rights abuses and environmental crimes, following submission of legal complaints by the Global Legal Action Network (GLAN). Pointing out that 336,000 people in La Guajira suffer from respiratory illnesses, the complaints quote a study which claims health has been damaged on a 'cellular' level due to the environmental changes wrought by the mine (GLAN, 2021, p. 1).

Latin America's biggest coalmine now faces a new level of scrutiny and test of its accountability – after almost 40 years of impunity. But the biggest question for communities in La Guajira is if the damage can ever be undone.

Sacred mountain

Today Cerrejón is synonymous with the 270-square-mile territory dominated since the mid-1980s by the mine, a former Exxon subsidiary that until July 2021 was jointly owned by BHP, Anglo American, and Glencore.

Photo 4.2 The Cerrejón coal mine / Hour.poing 2013 / CC BY-SA 3.0

But for centuries, the mountain of El Cerrejón was a sacred place where the Wayúu collected herbs for food and medicine. Water was always scarce on these rocky plains, but communities scraped a living from subsistence farming. About 45 per cent of La Guajira's population is indigenous; this includes around 98 per cent of Colombia's Wayúu population.

According to José Alvear Restrepo, the legal aid NGO which represents Provincial, more than 35 communities have been displaced by Cerrejón (Colectivo de Abogados José Alvear Restrepo, 2019, p. 6). In 1981, around 750 residents of the Wayúu village of Media Luna were forcibly evicted to make way for the construction of Puerto Bolívar on the Caribbean coast; today freight trains thunder along a 150 km railway line carrying coal to the port, from where it is shipped off around the world.

In 2001, Cerrejón made international headlines when bulldozers were brought in to demolish Tabaco and Afro-Colombian villagers were forced out by police, armed guards, and soldiers on behalf of the mine. Two decades later, the people of Tabaco are still fighting for justice; a Supreme Court ruling ordering the village to be rebuilt was never carried through.

'We've been a victim of community rights violations by multinationals and the state itself, which has permitted violations to occur. This has been a constant fight for our land, something which is the foundation for the survival of black communities,' says Rogelio Ustate.

Disconnection from their ancestral land is a spiritual loss for the people of Tabaco, as for the Wayúu. No amount of money could compensate those who have been scattered and cut off, and whose sacred places, such as mountains, lakes, rivers, and cemeteries, have been forever altered.

'For us our land is everything, a god,' says Ustate. 'We call it "Mother Earth" because it has nourished us ever since we came into being ... Our territory provided everything we needed: freedom, a stable economy based on agriculture and fishing, and free movement, which meant direct access to medicinal plants and water, something we no longer have.'

For the communities which have remained, the impact on their social, cultural, and spiritual life has been immense. Not only do they have to contend with pollution of the air, soil and water, the noise of freight trains and constant blasting make it hard for communities to uphold the spiritual and cultural practices of their ancestors. Cerrejón operates twenty-four hours a day, seven days a week.

'There is ... a sociocultural impact because we can't practise our customs and traditions due to their presence,' says Marcos Brito Uriana, a Wayúu leader from Provincial. 'For example, in our culture when someone has a dream it's like a premonition. We can avoid tragedies and many things through a dreamer's visions. But now they can't do that because they can't get a peaceful night's sleep.'

In 2017, Colombia's Constitutional Court recognized that displacement of the Tabaco community should not 'be treated exclusively as a problem of compensation' (Corte Constitucional de Colombia, 2017). In any case, compensation is little more than a token gesture for people whose losses are such that critics – including from the Center for Research and Popular Education (CINEP), a Colombian non-profit, and CAFOD, an international development charity linked to the Catholic Church – argue the abuses related to mining in La Guajira may amount to ethnocide and ecocide (Comunicaciones CINEP/PPP, 2020; CAFOD, 2021).

Dependency and distrust

'Because the state is distant and there's a lot of corruption at the departmental and municipal levels, it is a very unstable political system and in many ways the company has replaced the state,' says Banks. This situation has led to 'a lot of dependency between the communities and the company. People who have been resettled are dependent on them; they need some sort of job and need the company for everything' (see Chapter 3 for further examples).

This means the stakes in resettled communities are high, because residents are split as to whether to cooperate with *la mina*.

'Most of the divisions are in the resettled communities. In Provincial they basically bought out the leaders,' says Banks. 'I am not critical of them because they are thinking: "What can I do? I can't fight this monster."'

Indeed, communities can't agree on how to engage with what some call 'the huge enemy'. Luz Ángela Uriana says speaking out has led some people to ostracize her family.

'It has been really difficult for us. Me and my husband have no work and some people have cut us off. According to our customs, women are supposed to stay at home and people don't like to see women taking a leading role in the struggle.'

'If we were united rather than divided, we could do so much more,' echoes Brito Uriana. 'We would be able to force Cerrejón to accept the demands we present, if we were a united force with strength, but if we're divided and some people want this while others want that, then we won't get anywhere.'

In 2020 Cerrejón said it had reached a deal with the Wayúu in Provincial on implementing mitigation measures following the latest Constitutional Court ruling in December 2019. But indigenous leaders denied there had been a deal, accusing Cerrejón of a disinformation campaign designed to mislead national and international institutions and the public. According to the Provincial leaders, Cerrejón had attempted to divide the community by negotiating only with one leader and his family (Colectivo de Abogados José Alvear Restrepo, 2020).

It has been a similar story in Tabaco, where a small group of leaders signed a deal with the mine. However, this was never ratified by the community.

'The community is quite divided,' says Banks. 'There are people who want to sue the company, but the leadership wants to work with the company; the company makes deals with some people in the community and then says that this is taken care of.'

Distrust has grown not only as Cerrejón has divided communities, but as the Colombian state has repeatedly permitted the company to ignore court rulings. According to ABColombia, a coalition of British and Irish aid agencies, the state has failed to enforce multiple rulings aimed at addressing problems and upholding environmental guidelines (ABColombia, 2020).

'The issue is always enforcement,' says Banks. 'This has been an issue since the beginning – the Constitutional Court will make the order and the Ministry of Environment and regional environmental protection agency will always help the company at the end of the day. The orders are strong but how these are translated varies – weak interpretation and enforcement, and that is a major frustration for people in La Guajira.'

'The reality is the state lends a hand to the multinationals. It wears their uniform,' complains Ustate.

He argues that the Colombian state has not simply allowed Cerrejón to ignore court rulings, but has 'disrespected the Constitution' by giving the company permission to explore territories without even visiting to ascertain whether these were occupied by ethnic groups entitled to prior consultation.

Few issues are as contentious as water, a scarce resource of spiritual significance in La Guajira, Colombia's most arid region. According to the legal aid NGO José Alvear Restrepo, Cerrejón consumes 24 million litres of water a day (Colectivo de Abogados José Alvear Restrepo, 2019, p. 9).

There is only one river of significance, the Río Ranchería, now polluted after decades of mining activity in the department. Cerrejón also diverted part of the Arroyo Bruno – a tributary of the Río Ranchería – to mine the coal beneath it, thus endangering water supply for several Wayúu and Afro-Colombian communities.

Three Wayúu communities filed a legal challenge and in 2017 the Constitutional Court suspended work on the diversion, with the final decision to depend on the outcome of a study by a working group with input from the affected communities. However, the communities complain that their participation has been limited, and that the report produced by the working group depends mostly on information provided by Cerrejón. In April 2022 the Colombian government announced it intended to endorse the report, which may mean the diversion of the Arroyo Bruno remains in place (infobae, 2022).

'We don't have water and we don't have any access to water, says Uriana. 'The [court] judgment says they must provide 1,000 litres per family every two weeks, but for some families that is not enough. We have a micro aqueduct, but the water is so polluted it kills the animals. The pollution has not just affected the water, it has also destroyed the soil, which used to be fertile. The company is manipulative and they promise things that they don't deliver.'

The Wayúu say they are slowly dying of thirst. And yet, in La Guajira, those who dare to speak out may face harassment, criminalization or even death for their troubles.

War and peace

During Colombia's long civil conflict (1964-2016), a number of shadowy paramilitary groups carried out forced displacement, sexual violence, and assassinations of activists, labour leaders, indigenous people, journalists, teachers, and others.

Set up ostensibly to combat left-wing guerrilla groups such as the Revolutionary Armed Forces of Colombia (FARC) and the National Liberation Army (ELN), the paramilitaries often worked closely with large landowners, drug traffickers, politicians, businessmen, and multinational corporations, as well as the Colombian security forces. A 2013 report by the National Centre of Historical Memory, a government-created commission, blamed the paramilitaries for most of the killings during the conflict (Associated Press, 2013).

Many of those targeted were defending their land, and in recent decades this violence has increasingly been connected to the advance of mining and hydrocarbon extraction. The November 2016 peace deal between the government and FARC rebels did little to end this bloodshed. According to research by the NGO Front Line Defenders, Colombia has led the world ranking for killings of human rights activists since at least 2014, and the problem is getting worse, with more than 100 such killings reported every year from 2018 to 2021.

Significantly, the Business and Human Rights Resource Centre found that 44 per cent of attacks from 2015 to 2019 linked to businesses were against defenders who raised concerns about just five companies – one of which was Cerrejón (Business & Human Rights Resource Centre, 2020, p. 1).

Cerrejón has repeatedly condemned death threats directed towards its opponents – such as those sent to the Wayúu Women's Force by a group known as the Black Eagles in March and April 2020 – and denies any involvement. Still, human rights groups argue that Cerrejón has inflamed animosity towards activists, with company representatives having made a series of statements in the media attacking environmental and human rights defenders in La Guajira as obstacles to Colombia's development (Holland, 2020, p. 28).

In the neighbouring department of Cesar, the links between coal mining and paramilitary violence are even starker. Ex-paramilitaries and former contractors to Drummond and Prodeco have made statements under oath revealing how these two companies channelled funds to the United Self-Defense Forces of Colombia (AUC), a notorious paramilitary group active in the 1990s and 2000s and a direct predecessor of the Black Eagles (Moor and van de Sandt, 2014, pp. 9–10). In December 2020, the Colombian Attorney General charged the current and former presidents of Drummond's Colombian operations with conspiracy to finance paramilitary groups.

Though paramilitary violence remains a serious threat, Wayúu and Afro-Colombian leaders continue to resist. Over the decades, they have developed their communities' capacity to fight back peacefully, holding marches and protests and inviting national and international organiza-tions to visit, as well taking part in national forums and working with lawyers to pursue Cerrejón through the courts.

These efforts are becoming ever more critical, given that Cerrejón is coming towards the end of its life: its concession expires in 2033. Moreover, global demand for coal fell 4 per cent in 2020 and there is increasing pressure on companies to divest from the industry. In 2021 both BHP and Anglo American sold their 33.3 per cent interest in Cerrejón to partner Glencore in an effort to reduce their involvement in thermal coal, though critics argue it is a case of 'cut and run', with the companies trying to dodge accountability for the destruction caused by the mine.

'There is a big question mark over what happens when the mine closes in 2033,' says Emma Banks. 'They [the company] will take out what is left, and coal might not be profitable in a decade so what industry is going to replace it? If there is a strike, the entire economy in the area shuts down and no money changes hands – the dependency on the mine is on so many levels. This mine has been there for almost 40 years but what happens when it goes?'

In Provincial, Marcos Brito Uriana has focused on communications and intends to pass on these skills to younger people as a way to prepare them for the future.

'We have to create sustainable lifestyles for when Cerrejón goes. There are many people who depend on Cerrejón. So ... what are they going to rely on then?' he asks. 'It will be difficult to deal with what they will leave behind. We're never going to get the land back. We're never going to get the territory back. The flora and fauna are never coming back, so I don't know what the answer is.'

Defending its human rights record, Cerrejón points to negotiations on resettling communities and the complaints office it set up in 2010. It also highlights the adoption in 2011 of the UN's Guiding Principles on Business and Human Rights – though critics accuse it of failing to comply.

In spite of everything, Brito Uriana remains hopeful for the future.

'First and foremost, what I want is for the community to stay intact when Cerrejón goes ... with good housing, good access to roads, drinkable water, good education, a good economy and to expand the territory a little,' he says. 'Mostly, I want ways to guarantee a good quality of life.'

III. Cajamarca, Peru: police for hire

For Yeni Cojal Rojas, memories are still fresh of the events of 29 November 2011.

'Nineteen brothers were seriously injured, some in the face. Others were blinded. Our *compañero* Elmer Campos was paralyzed and they broke our *compañero* Carlos' leg. Shots were fired ...' she says, recalling the confrontation with police hired by Yanacocha, a mining company then owned by the American company Newmont, Peru's Buenaventura and the International Finance Corporation, the private lending arm of the World Bank.

Cojal Rojas is a *rondera* – a member of the peasant brigades which patrol rural areas in Peru – and belongs to the group Guardianes de las Lagunas. Alongside hundreds of other protestors, she had turned out to defend four mountain lakes in the department of Cajamarca.

But the protestors were met with tear gas, rubber bullets, and live ammunition. Campos, a local farmer, was shot in the back, losing his spleen and one of his kidneys, as well as being paralyzed from the waist down. He was just one of at least 24 protestors who were injured by police that day. That November marked the start of a brutal crackdown on opposition to Yanacocha that would continue into 2012 and beyond.

Since the 1990s, Yanacocha has operated a goldmine of the same name in Cajamarca which is one of the largest in Latin America. The company wanted to remove the lakes to make way for Conga, a massive new gold and copper project located just 24 kilometres north-east of the original mine. With a planned investment of $4.8 billion, it would have been the biggest mining investment to date in Peru.

Though Yanacocha claimed it would simply move the water into four reservoirs, locals believed this would destroy and contaminate water supplies, causing shortages for their livestock and crops. Their fears were understandable; in 2000 one of Yanacocha's contractors had spilled 150 kg of mercury, which is used to process gold, along a 43 km stretch of road around the village of Choropampa. People there say they have suffered ill health ever since (see Chapter 6 for more on the effects of mercury poisoning).

In February 2012, opponents of Conga travelled to Lima, for Peru's first ever 'National March for Water'. Some made the journey of more than 1,000 km on foot. In the capital they gathered with demonstrators from other mining-affected communities elsewhere in the country, for three days of rallies and events. Many of those who participated were women.

But these women were not spared the violence that was escalating in the highlands. 'The fight against Conga made us realize ... that there is no respect for women,' says Cojal Rojas. 'Because even though we women put ourselves forward to fight, [we] thought the police and army were not going to beat us. But the army beat us a lot ... and the police injured many women; they prosecuted many women.'

The Conga conflict led President Ollanta Humala to impose a state of emergency in Cajamarca in December 2011 and again in July 2012, giving police wide-ranging powers to crush the protests and bringing in the military to assist them. The bloodiest week of the conflict came in early July that year, when five people were shot dead by police, including one 16-year-old student (Cabitza, 2012; Grufides, 2016).

The following month, Newmont was forced to suspend Conga, and in 2016, with the project still mired in controversy, the company quietly downgraded its reserves to resources (reserves are resources which are considered viable to extract), admitting it 'did not anticipate being able to develop Conga for the foreseeable future' (Jamasmie, 2016).

But for Cojal Rojas, and the people of Cajamarca, Conga's suspension signified not the end of the conflict, but a 'truce' – another staging post in their never-ending struggle to protect their land.

Gold rush

The city of Cajamarca is known as the place where the last Inca emperor, Atahualpa, filled a room with gold to pay the ransom for his release, following his capture by the *conquistadores* in 1532. Atahualpa was as good as his word and handed over the gold – but the Spaniards killed him anyway.

As of 2021, 42.5 per cent of the population of the department of Cajamarca was living in poverty, despite nearly three decades of large-scale mining (Instituto Peruano de Economía, 2021). During this time mining has brought social upheaval, state appropriation of land and drastic consequences for the health of people and the environment.

'There is heavy metal poisoning throughout almost the whole Cajamarca region, says Cojal Rojas. 'Every day people have to take eight or nine pills just to be able to sustain themselves. They are dying little by little.'

As in neighbouring Colombia, in Peru mining projects have been systematically prioritized over the protection of human rights – and repression of those who protest has been brutal. A 2014 report by Global Witness found that Peru was the world's fourth most dangerous country to be an environmental defender, with at least 57 killings between 2002 and 2014 (Global Witness, 2014, p. 3).

The violence surrounding Conga was also accompanied by more than 300 criminal charges filed against protestors, encouraged by a complicit national media which stigmatized the protests and smeared activists as drug traffickers and terrorists.

'There weren't just attacks by snipers, but a violent repression in terms of arrests. Anyone who was involved in the protest was detained indiscriminately; today they are the people who were criminalized in the Conga case,' says Cojal Rojas.

Cojal Rojas was among 16 human rights defenders who were criminalized for their role in the protest against the Conga mine, with a prosecutor seeking more than 30 years' imprisonment for each of them. On 28 March 2017 the Supreme Court of the Supra Provincial of Cajamarca finally found them innocent.

But the interminable hearings took a toll on Cojal Rojas, who was pregnant with twins during that time. She lost her baby daughter in childbirth and her son was born with high blood pressure and other health problems related to the stress she had suffered.

'That is violence not only against my rights as a woman but also as a mother,' she says.

In July 2020 a report by EarthRights International found that thousands of human rights defenders had suffered harassment, criminalization, violence, and intimidation. Efforts to seek redress in the courts were subject to long delays, while many were slapped with trumped-up charges or made to travel long distance for hearings, it said (EarthRights International, 2020, p. 4).

One case of harassment in Cajamarca that did make headlines outside Peru is still the subject of a lawsuit against Newmont, over a decade later. Máxima Acuña de Chaupe, a potato farmer who refused to sell her family's 60 acres to Yanacocha, recalled how in 2011 police entered her home in Cajamarca and assaulted her children.

'I was grabbed by six policemen, three on each arm grabbed me from behind and beat me with batons. Then they threw me to the ground and beat my son, who was taking photographs, on the arms and chest,' she told *The Guardian*. 'The special forces hit my daughter in the head with the butt of the machine gun. Four of them cornered my youngest son and pointed their machine guns at him, warning him not to shout, not to call out, not to run' (Collyns, 2016).

Since Yanacocha began exploration work for Conga in 2011 – without conducting prior consultation with local communities beforehand (Coordinadora Nacional de Derechos Humanos, 2019) – the David and Goliath struggle between Acuña, Yanacocha and its parent companies has seen the potato farmer not only beaten and harassed, but also dragged through the courts.

Yanacocha brought a case against Acuña and her family in 2011 for the crime of 'aggravated encroachment' – in other words, they were accused of illegally occupying the land they had lived on since the 1990s. A provincial court sided with Yanacocha in August 2014, handing Acuña a suspended prison sentence of two years and eight months and a fine of nearly $2,000 – a fortune for a subsistence farmer in Peru.

Photo 4.3 Máxima Acuña on her farm in the highlands of Cajamarca / © Goldman Environmental Prize 2015

But Acuña held firm, with national and international support, managing to get the verdict overturned by a court of appeal in December 2014. Newmont tried to have this reversed, but in 2017 the Peruvian Supreme Court declared Acuña innocent of all charges. For her determined stand she won the Goldman Environmental Prize in 2016 and became an international symbol of resistance.

In 2017 Acuña and her family took the fight to the United States, suing Newmont in Delaware for damages. Alleging harassment and physical and psychological abuse, Acuña's lawyers argued that they would not get a fair hearing in Peru due to judicial corruption and Newmont's ongoing influence over the government and courts.

But while the Acuña de Chaupe family and their allies are hoping to set an important precedent, courts in countries of the Global North are usually extremely reluctant to hear cases on human rights abuses committed by companies overseas, arguing that the responsibility lies with the local courts (see Chapter 2). Indeed, after several years of legal wrangling, it was ruled that *Acuña-Atalaya v. Newmont Mining Corp* would not be heard in the United States.

Wyatt Gjullin, a lawyer with Earthrights, says that in most land disputes in Peru such abuse goes unpunished.

'In Latin America there isn't a culture of holding companies accountable; a lot of human rights lawyers try to hold the state accountable and there are more opportunities to do this,' he says. 'Peru has many environmental laws that are not enforced, but research has shown that a high percentage of environmental infractions are just forgiven by the courts.'

History of conflict

Though Peru is a country of long mining history, the industry has expanded aggressively since the early 1990s, leading to a proliferation of social conflicts all over the country. Over the years, communities have become increasingly savvy and well organized in defence of their lands.

José De Echave is co-founder of the rights organization CooperAcción and a former vice-minister of environmental management at Peru's Ministry of the Environment. He resigned from the government in late 2011 over its handling of the Conga conflict and other socioenvironmental clashes elsewhere in the country.

'Thirty years ago, the conflict was mainly about labour rights but today the focus has shifted to a struggle over territory as minerals are located close to indigenous lands,' he says. 'Few people heard about those earlier struggles, but land and environmental defenders have much higher visibility now.'

In cities far from the affected areas, this unrest has often been misrepresented and misunderstood. Nonetheless, 'What people are really fighting for is their rights,' he insists.

This gulf between Peru's coastal cities and its rural, highland interior was revealed in stark fashion by the 2021 presidential election. Pedro Castillo, the eventual winner, is the son of illiterate campesinos from Cajamarca and a former *rondero*. He owes his victory to near-unanimous support in similar impoverished, mining-affected regions throughout the country. Edging out rival Keiko Fujimori by just 44,000 votes in the runoff, Castillo's share of the vote was over 90 per cent in provinces such as Cotobambas, Espinar and Chumbivilcas, all home to some of Peru's biggest mines (Reuters, 2021).

In his inaugural speech, Castillo alluded to the 'millennial mining and agricultural tradition' and vowed to 'put order in mining', distributing profits more equally and eradicating corruption (gob.pe, 2021, p. 8; p. 15). There were also promises on the environment, such as a new law to order territory and a better system for monitoring environmental infractions. While Castillo's victory provoked horror amongst Peru's political and business elite, such pledges resonated in regions that have long been neglected and in recent years, increasingly handed over to mining companies with the blessing of the state.

This has included providing state security forces, not only to police conflicts when they erupt, but even to work as company security. A series of laws have been passed since the 1990s permitting private companies to contract services from the Peruvian National Police (PNP); mining companies have been some of the primary beneficiaries of this arrangement (EarthRights International et al., 2019, p. 9). During this time, many egregious human rights abuses linked to the industry have been committed by the PNP, hired as security in areas of mining conflict.

'The police are an institution of the state of Peru that is supposed to guarantee security,' says De Echave. 'But what happens in the zones where there is influence of mining companies [is that] the police sign agreements with the companies, so the national police become a private police force of the companies and cease to carry out their function.'

'What single piece of legislation would improve the situation for land and human rights defenders in Peru?' asks Wyatt Gjullin. 'Abolish the legal framework that permits these agreements between private companies and the police.'

Throughout these conflicts, a major challenge for Peru's environmental defenders has been the lack of national laws through which to assert their human rights. But they have developed other successful strategies, including occupations of mine sites and nearby areas, and building capacity, such as forming groups to test water for contamination.

Women will always be at the heart of this struggle to defend the environment, says Yeni Cojal Rojas: 'The violence [of the mining industry] violates the rights of women more because women are in contact with water all the time, every day, for food, and for their family's hygiene. Women are more affected because for us the priorities are our children, and keeping a clean and dignified home.'

'Let us not be opponents of one issue, but rather let us fight together against a greater power, which is the model of the world system,' she says.

***

Today, Latin America is the epicentre of human rights abuses connected to extractive industries. This is the result of an extended historical period in which governments of all political persuasions have seen the expansion of mining and hydrocarbons as a key engine of development and source of revenue.

Laws to attract foreign investment and facilitate extraction on the one hand have been accompanied by weak governance and corruption on the other. This has left communities at the mercy of powerful corporations that exploit their land for profit, with scant regard for their rights to life, livelihoods, water, and food. Indigenous communities and other marginalized groups are particularly vulnerable.

Those who fight back suffer stigmatization, harassment, criminalization, and violent repression. Women involved in these struggles are not spared; in some cases, they are even targeted specifically. The violence may come from the police or military, private security guards employed by mining companies, or criminal and para-state groups. Hundreds of thousands have been forced to flee home in search of safety – and a way to survive.

Though communities seek redress through national courts and international forums, rulings in their favour often go unenforced and environmental infractions unpunished. Nation states across Latin America have shown a systematic lack of political will when it comes to upholding the human rights of communities involved in mining conflicts.

The positive news is that communities have been fighting back – locally, nationally, and internationally – to demand that governments, intergovernmental organizations and global investors keep tabs on companies and hold them to account whenever they fail to meet their human rights obligations. This has involved travelling to countries where the companies are headquartered and where the minerals and metals extracted are consumed.

'We want people in the countries which buy this coal to think more in depth about where it comes from,' says Marcos Brito Uriana. 'What effect does it have in the country or region where it's mined?'

References

All references to web-based material were checked and still available in November 2022 unless otherwise stated.

All references are listed, with clickable links for your convenience, on the page for this chapter on the Heart of Our Earth website: <https://lab.org.uk/the-heart-of-our-earth/>

ABColombia (2020) 'ABColombia profound concerns regarding Cerrejon's lack of compliance with Court Rulings'. [online] Available at: <https://www.abcolombia.org.uk/cerrejon-lack-of-compliance-with-court-rulings/>.

Associated Press (2013) 'Colombian conflict has killed 220,000 in 55 years, commission finds'. [online] *The Guardian*. Available at: <https://www.theguardian.com/world/2013/jul/25/colombia-conflict-death-toll-commission>.

BBC News (2022) 'Juan Orlando Hernández: Honduran ex-leader extradited to US'. [online] Available at: <https://www.bbc.co.uk/news/world-latin-america-61174692>.

Beeton, D. (2012) 'Miguel Facusse is Tragically Misunderstood'. [online] Center for Economic and Policy Research. Available at: <https://cepr.net/honduras-most-powerful-man-is-simply-misunderstood-qi-probably-had-reasons-to-kill-himbut-im-not/>.

Business & Human Rights Resource Centre (2020) *Business & Human Rights Defenders in Colombia*. [online] Available at: <https://media.business-humanrights.org/media/documents/files/Business__Human_Rights_Defenders_in_Colombia.pdf>.

Cabitza, M. (2012) 'Peru mine disputes mar President Humala's first year'. [online] BBC News. Available at: <https://www.bbc.co.uk/news/world-latin-america-18980109>.

CAFOD (2021) *Protecting our common home: land and environmental human rights defenders in Latin America* [online] CAFOD. Available at: <https://cafod.org.uk/content/download/56617/776987/version/2/file/Protecting%20our%20common%20home%20HDR%20in%20Latin%20America_v5.pdf>.

Call, C. (2021) 'The imperative to address the root causes of migration from Central America'. [online] Brookings. Available at: <https://www.

brookings.edu/blog/order-from-chaos/2021/01/29/the-imperative-to-address-the-root-causes-of-migration-from-central-america/>.

Chavkin, S. (2015) 'Tierras bañadas en sangre'. [online] *El País*. Available at: <https://elpais.com/elpais/2015/06/08/planeta_futuro/1433761148_555067.html>.

Colectivo de Abogados José Alvear Restrepo (2019) *Diez verdades sobre Carbones de Cerrejón*. [online] Available at: <https://www.colectivodeabogados.org/old/IMG/pdf/diez_verdades_sobre_carbones_de_cerrejon.pdf>.

Colectivo de Abogados José Alvear Restrepo (2020) 'Carbones del Cerrejón miente y actúa de manera fraudulenta frente a sentencia judicial y respuesta a los Relatores Especiales de Naciones Unidas'. [online] Available at: <https://www.colectivodeabogados.org/old/?Carbones-del-Cerrejon-miente-y-actua-de-manera-fraudulenta>.

Collyns, D. (2016) 'Goldman prize winner: 'I will never be defeated by the mining companies''. [online] *The Guardian*. Available at: <https://www.theguardian.com/environment/2016/apr/19/goldman-prize-winner-i-will-never-be-defeated-by-the-mining-companies>.

Comunicaciones CINEP/PPP (2020) 'Noche y Niebla N°61: Minería de Carbón y Des-Arroyo'. [online] CINEP. Available at: <https://www.cinep.org.co/es/mineria-de-carbon-y-des-arroyo/>.

Coordinadora Nacional de Derechos Humanos (2019) 'Consulta previa para Estudios de Impacto Ambiental'. [online] Available at: <https://derechoshumanos.pe/2019/11/consulta-previa-para-estudios-de-impacto-ambiental/>.

Corte Constitucional de Colombia (2017) 'Sentencia T-329/17'. [online] Available at: <https://www.corteconstitucional.gov.co/relatoria/2017/T-329-17.htm>.

Davies, L. (2021) 'Land defenders: will the Cáceres verdict break the 'cycle of violence' in Honduras?'. [online] *The Guardian*. Available at: <https://www.theguardian.com/global-development/2021/jul/15/will-caceres-verdict-help-end-honduras-cycle-of-violence>.

Defensores en Línea (2019) 'La Tolva mecanismo de castigo del régimen para los opositores políticos'. [online] Available at: <https://www.defensoresenlinea.com/la-tolva-mecanismo-de-castigo-del-regimen-para-los-opositores-politicos/>.

Dinero HN (2022) 'Honduras se declaró libre de minería a cielo abierto y cancelará concesiones: inversionistas en jaque'. [online] Available at: <https://dinero.hn/honduras-se-declaro-libre-de-mineria-a-cielo-abierto-y-cancelara-concesiones-inversionistas-en-jaque/>.

EarthRights International (2020) *Criminalization of Earth Rights Defenders in Peru. Highlights from the Report*. [online] Available at: <https://earthrights.org/wp-content/uploads/informe_criminalización_highlights_final-1.pdf>.

EarthRights International, Instituto de Defensa Legal and the Comisión Nacional de Derechos Humanos (2019) *Convenios entre la Polícia Nacional y las empresas extractivas en el Perú. Análisis de las relaciones que permiten la violación de derechos humanos y quiebran los principios del Estado democrático de Derecho*. [online] Available at: <https://earthrights.org/wp-content/uploads/Informe-Convenios-entre-PNP-y-empresas-extractivas.pdf>.

El Heraldo (2018) 'Investigan uso de AK-47 en ataque a los dos militares muertos en Tocoa.' [online] Available at: <https://www.elheraldo.hn/sucesos/investigan-uso-de-ak-47-en-ataque-a-los-dos-militares-muertos-en-JVEH1229415>.

Global Legal Action Network (2021) *Non-compliance with the OECD guidelines for multinational enterprises. BHP, Anglo American, and Glencore.* [online] Available at: <https://www.glanlaw.org/_files/ugd/14ee1a_14d758179e494 a7bb096875cf1f63c87.pdf>.

Global Witness (2014) *Peru's Deadly Environment. The Rise in Killings of Environmental and Land Defenders.* [online] Available at: <https://www. globalwitness.org/documents/10510/peru_deadly_environment_en.pdf>.

Global Witness (2017) *Honduras: The Deadliest Place to Defend the Planet.* [online] Available at: <https://www.globalwitness.org/documents/18804/ English_Honduras_full_report_single_v6.pdf>.

gob.pe (2022) *Discurso de asunción del Presidente de la República, José Pedro Castillo Terrones, 28 de julio de 2021.* [online] Available at: <https://cdn.www. gob.pe/uploads/document/file/2049663/Mensaje_a_la_nacion_presidente_ Pedro_Castillo.pdf.pdf>.

Gordon, T. and Webber, J. (2016) *Blood of Extraction. Canadian imperialism in Latin America.* Halifax, Nova Scotia and Winnipeg, Manitoba: Fernwood Publishing.

Grufides (2016) 'Conga: Muertes ocurridas durante el conflicto siguen impunes'. [online] Available at: <http://grufides.org/blog/conga-muertes-ocurridas-durante-el-conflicto-siguen-impunes>.

Hammond, B., Berghoef, M., Ferrucci, G., Grzyb, A., Lascaris, D. and Montoya, A. (2020) *Guapinol Resiste. Orígenes del conflicto minero en el Bajo Aguán, Honduras.* [online] Available at: <https://www.acafremin.org/images/ documentos/Guapinol_ESP_Baja_Res.pdf>.

Holland, L. (2015) 'The Dangerous Path Toward Mining Law Reform in Honduras'. [online] Council on Hemispheric Affairs. Available at: <https://www.coha. org/the-dangerous-path-toward-mining-law-reform-in-honduras/#_ftn6>.

Holland, L. (2020) *Undermining Human Rights: Ireland, the ESB and Cerrejón coal.* [online] Christian Aid Ireland. Available at: <https://www.christianaid.ie/sites/ default/files/2020-02/Cerrejon%20Report.pdf>.

Human Rights Watch (2020) 'Colombia: Indigenous Kids at Risk of Malnutrition, Death'. [online] Available at: <https://www.hrw.org/ news/2020/08/13/colombia-indigenous-kids-risk-malnutrition-death>.

Human Rights Watch (2021) 'World Report 2021: Honduras'. [online] Available at: <https://www.hrw.org/world-report/2021/country-chapters/honduras>.

infobae (2022) 'Gobierno nacional avala la destrucción del arroyo Bruno, según comunidades de La Guajira'. [online] Available at: <https://www. infobae.com/america/colombia/2022/04/08/gobierno-nacional-avala-la-destruccion-del-arroyo-bruno-segun-comunidades-de-la-guajira/>.

Institute of Latin American Studies (2020) '2012–2013 Creation and reconfigu-ration of Montaña de Botaderos National Park'. [online] Available at: <https:// ilas.sas.ac.uk/research-projects/legal-cultures-subsoil/2012-2013-creation-and-reconfiguration-montaña-de-botaderos>.

Institute of Latin American Studies (2022) '2018–2022 Criminalisation and detention of CMDBCP environmental rights defenders'. [online] Available at: <https://ilas.sas.ac.uk/research-projects/legal-cultures-subsoil/2018-criminalisation-and-detention-cmdbcp-environmental>.

Instituto Peruano de Economía (2021) 'Cajamarca: la quinta región más pobre de 2020'. [online] Available at: <https://www.ipe.org.pe/portal/cajamarca-la-quinta-region-mas-pobre-de-2020/>.

Jamasmie, C. (2016) 'Community opposition forces Newmont to abandon Conga project in Peru'. [online] Mining.com. Available at: <https://www.mining.com/community-opposition-forces-newmont-abandon-conga-project-peru/>.

Lakhani, N. (2016) 'Remembering Berta Cáceres: 'I'm a human rights fighter and I won't give up''. [online] *The Guardian*. Available at: <https://www.theguardian.com/world/2016/mar/03/remembering-berta-caceres-interview-la-esperanza-honduras-human-rights>.

Moor, M. and van de Sandt, J. (2014) *The Dark Side of Coal. Paramilitary Violence in the Mining Region of Cesar, Colombia*. [online] PAX. Available at: <https://paxvoorvrede.nl/media/download/pax-dark-side-of-coal-final-version-web.pdf>.

Office of the United Nations High Commissioner for Human Rights (2016) 'Berta Cáceres' murder: UN experts renew call to Honduras to end impunity'. [online] Available at: <https://www.ohchr.org/en/press-releases/2016/04/berta-caceres-murder-un-experts-renew-call-honduras-end-impunity?LangID=E&NewsID=19805>.

Office of the United Nations High Commissioner for Human Rights (2018) 'Honduras election protests met with excessive and lethal force – UN report'. [online] Available at: <https://www.ohchr.org/en/press-releases/2018/03/honduras-election-protests-met-excessive-and-lethal-force-un-report?LangID=E&NewsID=22799>.

Office of the United Nations High Commissioner for Human Rights (2019) 'Statement at the end of visit to Honduras by the United Nations'. [online] Available at: <https://www.ohchr.org/en/statements/2019/08/statement-end-visit-honduras-united-nations?LangID=E&NewsID=24925>.

Office of the United Nations High Commissioner for Human Rights (2020) 'UN expert calls for halt to mining at controversial Colombia site'. [online] Available at: <https://www.ohchr.org/en/press-releases/2020/09/un-expert-calls-halt-mining-controversial-colombia-site>.

Office of the United Nations High Commissioner for Human Rights (2021) *Opiniones aprobadas por el Grupo de Trabajo sobre la Detención Arbitraria en su 89o período de sesiones, 23 a 27 de noviembre de 2020*. [online] Available at: <https://www.ohchr.org/sites/default/files/Documents/Issues/Detention/Opinions/Session89/A_HRC_WGAD_2020_85.pdf>.

Olson, J. (2020) 'Honduras's Deadly Water Wars'. [online] *The Nation*. Available at: <https://www.thenation.com/article/world/honduras-mine-conflict/>.

PBI Honduras (n.d.) 'La otra cara de la militarización: vulneraciones de derechos humanos'. [online] Available at: <https://pbi-honduras.org/es/news/2021-09/la-otra-cara-de-la-militarización-vulneraciones-de-derechos-humanos>.

Reuters (2021) 'Elecciones en Perú: Pedro Castillo arrasa en las regiones mineras clave'. [online] *El Economista*. Available at: <https://web.archive.org/web/20220131080137/https://www.eleconomista.com.mx/internacionales/Elecciones-en-Peru-Pedro-Castillo-arrasa-en-las-regiones-mineras-clave-20210611-0038.html>.

University of Virginia School of Law (2020) *Pretrial monitoring of the Guapinol and San Pedro community*. [online] Available at: <https://ilas.sas.ac.uk/sites/default/files/reports/English%20UVA%20Report%20on%20Guapinol.pdf>.

CHAPTER 5
The mud is still flowing

This chapter examines three major mining disasters of recent years: a chemical spill at a copper mine, in Sonora, Mexico, in 2014; the Bento Rodrigues tailings dam disaster, in Minas Gerais, Brazil, in 2015; and the Brumadinho dam disaster, also in Minas Gerais, Brazil, in 2019. It looks at the failure to repair the damage caused and provide justice to people who lost homes, loved ones, and livelihoods. It argues that in the absence of any meaningful reform of the mining industry, and given that those responsible have enjoyed near total impunity, such disasters are likely to recur.

I. The Río Sonora spill, Mexico

'After almost seven years – it's been almost seven years! – there is still no authority that can tell us with certainty that the water from our taps is fit for human consumption,' says Ramón Miranda, a shopkeeper from Aconchi, in the state of Sonora, in the far northwest of Mexico. The exasperation in his voice is palpable.

On 6 August 2014, there was a chemical spill at the Buenavista de Cobre copper mine, 128 km to the north of Aconchi. Dubbed 'the worst disaster in the history of the Mexican mining industry' by the then Minister of the Environment, Juan José Guerra (Servín, 2014), Buenavista spilt 40,000 m³ of a copper sulphate acid solution, which flowed first into the Río Bacanuchi, then the Río Sonora. Enough to fill 16 Olympic-size swimming pools, it stained a 60 km stretch of the Río Sonora a lurid shade of orange.

In the immediate aftermath of the spill, Buenavista de Cobre remained silent. The company let more than a day pass before notifying the authorities of the spill, during which time local people continued using their water as normal (PODER, 2015, p. 11).

'We learned on Facebook from the local authority in Arizpe that polluted water was flowing down the river. That's how we found out about this terrible spill,' says Miranda. Other residents of affected communities found out via word of mouth, or by witnessing first-hand the unnerving change in the colour of the water.

The company did not publicly recognize the disaster until six days after it occurred, claiming it was the result of 'above average rainfall,' (CCRS and PODER, 2020, p. 1) though in fact, the spill was caused by a failure on a pipe seal in mining installations that were unfinished and were operating illegally, without the necessary environmental permits (Blust, 2019).

According to Mexican health and environmental authorities, the spill affected 24,048 people across seven municipalities, from Arizpe in the north down to Ures in the south, not far from the state capital Hermosillo (Fundar, 2018, p. 66). In early 2015, affected people from all seven municipalities set up the Comités de Cuenca Río Sonora ('Sonora River Basin Committees') (CCRS), to demand that the government and the company remedy the situation.

But seven years later, they complain that very little has been resolved.

'Children are drinking the water'

The copper sulphate solution that flooded into the Bacanuchi and Sonora rivers contained heavy metals including nickel, iron, arsenic, lead, cadmium, manganese, and aluminium. Soon after the spill, people from affected communities began reporting aches and pains; itchy, irritated skin; even alopecia (Cárdenas, n.d., a).

Due to their long half-life, most heavy metals are excreted very slowly from the body. This means that if an individual is exposed to these substances over a period of months or years, they accumulate in the system and may cause severe, chronic, and in some cases irreversible health issues.

This is consistent with the official statistics: in September 2014, just a month after the disaster, 19 people with related health issues had been identified. A month later, it was 36. By the following March, that figure had spiralled to 270 and by August 2016, two years on from the spill, it stood at 381, at which point, the state authorities responsible for addressing the health impacts of the spill stopped publishing data (CCRS and PODER, 2018, p. 4). Today, there may be more than 1,000 people suffering health complications who have not been recognized by either the company or the authorities (CCRS and PODER, 2020, p. 6).

Indeed, locals report increased numbers of people suffering from renal problems and various types of cancer. But as systematic monitoring was never implemented, the true magnitude of the disaster may never be known. Meanwhile, people in these communities live in a state of perpetual anxiety. Since the spill, they have been forced to depend on bottled water, at least for drinking – an added cost to which people are not accustomed, given that prior to August 2014 most drank the tap water.

'The wells are in the same places and the local people continue to use this water, not to drink, but to wash and have a shower, which causes skin problems,' says Fernanda Hopenhaym, chief executive director at the non-profit PODER, which has worked closely with the communities affected by the spill. 'They also give it to their animals to drink and water their crops with it. So there is regular domestic use of the water from contaminated wells, even if they don't drink it. And sometimes, when they can't afford to buy drinking water, they have to drink the contaminated water, though they boil it.'[1]

'Buying bottled or purified water simply leaves people without money to pay for other needs,' echoes Miranda. 'People told us – and this is shocking – that instead of buying bottled water they preferred to buy beans, rice, or other food. What does that mean? People are drinking the water. Children are drinking the water.'

Grupo México calls the shots

Buenavista de Cobre is located in Cananea, a historic mining town just 60 km away from the border with the U.S. state of Arizona. It has been running since 1899, making it the oldest mine in constant operation in North America. The largest copper mine in Mexico and the fourth largest in the world by production, it is owned by the mining and logistics giant Grupo México, the CEO of which is the reclusive billionaire Germán Larrea. According to *Forbes*, Larrea was worth $28.1 billion in 2021, making him the second richest man in Mexico behind the telecoms magnate and investor Carlos Slim.

In September 2014, a month after the spill, the Mexican government and Grupo México set up the Fideicomiso Río Sonora (FRS) – a kind of trust fund – to repair the damage caused and attend to the needs of people affected, providing compensation where appropriate. Grupo México committed two billion pesos (approximately $154 million at the time) and promised to increase that figure if necessary, subject to the findings of a study on the impacts of the spill (Lamberti, 2018, pp. 5-6).

The study was never carried out, nor was there ever any serious attempt to calculate the damages and determine how much compensation was due to those affected in the Río Sonora communities. They complain that the compensation offered was arbitrary, opaque, and sometimes unequal (Cárdenas, n.d.,b). For example, each household the FRS identified as being affected was granted a one-off payment of 15,350 pesos (approximately $1,180 at the time) as compensation for the loss of their water supply.

'Studies we've done with PODER showed that the money lasted less than a year because of the spending on bottled water,' says Miranda.

The effect of the spill on the local economy has been devastating, given that the region depended heavily on agriculture. Prior to the spill there was cattle farming; people cultivated chillies, peanuts, alfalfa, and corn; and they made cheese and other local products. But many farmers have had to abandon the sector and those who have persisted complain that the results are not the same. Unsurprisingly, sales have fallen, with consumers now wary of the regional produce.

But local farmers complain that the compensation they have been offered is a pittance. 'I lost my entire production,' says Marco Antonio García Méndez. 'So back then I thought I'd get half a million pesos for it. What did they give me? 70,000 and a water tank' (Cárdenas, n.d.,b).

The responsibility for deciding how the funding of a *fideicomiso* is spent lies with a Technical Committee. In the case of the FRS, this consisted of two members appointed by the Ministry of the Environment (SEMARNAT), two appointed by Grupo México – one Grupo México engineer and a supposedly independent expert – and one further member appointed in an agreement between Grupo México and the government. The affected people in the Río Sonora region had no representation on the Technical Committee, nor were they invited to contribute in any meaningful way.

'The ones with the real power of decision are Grupo México,' says Hermes Ceniceros, spokesperson for the Fronteriza Health and Environment Network, a local NGO. 'This means that the *fideicomiso* works as a branch of the company, acting as an authority in the territory affected but without being accountable in the same way. In other words, as the *fideicomiso* is privately funded, they can deny any request for information about how the money is being used' (Fundar México, 2018).

Ultimately, the FRS spent only around 1.2 billion of its 2 billion-peso budget, a figure equivalent to around 0.00016 per cent of Grupo México's annual profits (Contreras, 2020). Of this 1.2 billion, just over half went directly on attending to the needs of affected communities. The rest of the money was granted to other recipients including Grupo México itself; Mexican government at federal, state, and municipal level; and government departments such as SEMARNAT and the Secretariat of Social Development. The reasons for these payments remain unknown (Lamberti, 2018, p.12; p. 16).

Just 7.8 million pesos were spent on attending to those whose health had been affected by the spill – four times less than the total spent on public relations. This included publicity slots about the environmental recovery of the river which were shown in the Cinemex cinema chain – another asset in Grupo México's portfolio (CCRS and PODER, 2021, p. 7).

Broken promises

This abject mismanagement of resources resulted in failure to deliver on two key promises.

In January 2015, Rodolfo Lacy Tamayo, president of the FRS, announced the construction of 36 water purification plants to eliminate heavy metals from communities' water supply. He promised these would be operational in the following seven months (CCRS and PODER, 2020, p. 2).

They were not. Grupo México reduced this number to 28 in 2016 and 9 a year later (Vega, 2019), claiming that they were unnecessary as levels of heavy metals in the river had fallen. In the end, only six were ever built, and out of these six, only two have ever been operational, and even then, only intermittently. The plants have been beset by technical problems, while municipal authorities in these remote communities have neither the financial resources nor the technical capacity to ensure continuous operation at the level required (Díaz-Caravantes et al., 2021, Table 6).

Aside from the one-off payment, all that residents received to mitigate the impact on their water supply was a water tank with a Fideicomiso Río Sonora label on the side. 10,188 of these tanks were installed outside local homes and businesses, supposedly to store the water that came from the treatment plants. For the first few months the FRS supplied drinking water to these tanks, but the arrangement soon petered out. Many of the tanks have since fallen into disuse, visible symbols of waste, neglect, and corruption.

Nearly 185 million pesos (approximately $14.2 million) were spent on these water tanks, at a cost of around 18,000 pesos per tank. According to estimates by journalists at the newspaper *Excelsior*, this figure was around three times the market value of the tanks and their installation. Of these 185 million pesos, 28.6 million were transferred to the Mexican College of Environmental Engineers, an organization founded by FRS president Rodolfo Lacy Tamayo, while a further 5.5 million went to the company Rotoplas, owned by Carlos Rojas Mota Velasco – a cousin of Germán Larrea who sits on the board of directors at Grupo México (Méndez and Sánchez Dórame, 2018).

The second key promise that the FRS reneged on was the Epidemiological and Environmental Monitoring Unit of Sonora (UVEAS). Announced by the Ministry of Health in October 2014, the UVEAS was to be 'a first-class clinic, with services provided by specialized staff and the most advanced technology' (COFEPRIS, 2014). It was to operate for 15 years, ensuring serious medium- to long-term monitoring of the health of communities affected by the spill and proper treatment for whoever might require it.

But Grupo México never provided enough funding for the project to be completed (Lamberti, 2018, p. 13). At the end of June 2016, the provisional UVEAS clinic was closed and its 17 health professionals all laid off. The three buildings belonging to what was to be the permanent UVEAS complex in Ures were never completed and in November 2019 Grupo México handed them over to the National Guard. As with the water treatment plants, the justification for the failure to deliver the clinic was that levels of heavy metals in the water had stabilized and so the project was unnecessary – though by this point, the health impacts had been well documented, with cases running at least into the hundreds.

Finally, after years of obfuscation by both Grupo México and the government, studies carried out since 2019 have confirmed the worst fears of residents of the Río Sonora basin: that the water across all seven affected municipalities is severely contaminated with heavy metals, principally arsenic and lead.

'What's it going to take for the authorities to do something?' asks Ramón Miranda. 'The evidence is clear; this new data just tells us what we already know. For the last six years, ever since the mess that Grupo México left behind, the water we've been drinking has been poisoning us, but the authorities are doing nothing about it.' (PODER, 2020)

" Que las autoridades
dejen de solapar
las atrocidades
de Grupo México.
Necesitamos
gobernantes
con memoria. **"**

@CuencaRíoSonora
@ProjectPODER

#RíoSonora6
#6AccionesParaElRío

Photo 5.1 The photo shows Ramón Miranda. The quotation reads 'The authorities must stop covering up abuses by Grupo México. We need leaders with memory.' / © CCRS/PODER 2021

Legal victories

In February 2017 the FRS was quietly closed by the government and Grupo México. This was done unilaterally and without even informing the people affected by the spill, who didn't find out until the decision emerged in the press six months later (Lamberti, 2018, p. 19).

'It just closed without finishing its job, having spent only 60 per cent of its funding,' says Fernanda Hopenhaym. 'There are 800 million pesos still left in that fund. They were never used and are presumably still held by the company.'

With the help of PODER, the Comités de Cuenca Río Sonora (CCRS) appealed against this decision, and in January 2020 the Mexican Supreme Court ruled in their favour, stating that there should be a proper process of consultation with those affected by the disaster, with the aim either of reopening the FRS, or establishing some other mechanism for remediation.

'This was an important victory – the most important one I think for the people affected – in terms of the prospects of genuine remediation,' says Hopenhaym.

During 2022, the CCRS attended a series of meetings with representatives from various government agencies. New plans were presented which aim to provide access to healthcare and clean water to those affected by the spill, alongside a revised analysis of the economic losses caused. These are welcome developments, but as the CCRS recognizes, this is just the beginning of a process of justice for which they have been fighting for nearly eight years.

The other major legal victory obtained by PODER and the CCRS relates to the community of Bacanuchi, a tiny hamlet of around 300 people in Arizpe, the northern-most municipality affected by the spill. In 2016 residents of Bacanuchi noticed that Grupo México was carrying out major construction work nearby. They soon realized that the work was the massive expansion of a tailings dam, just 20 km away from the community. With residents still suffering from the effects of the 2014 spill, this news was obviously cause for concern.

'If there were a failure, it would wipe the community off the map,' says Hopenhaym. 'It would be a disaster even worse than the spill in 2014. The dam will contain 51,000 times more in volume than the 40 million litres that were spilt … So, just imagine. It's enormous.'

With PODER's assistance, local residents filed an injunction against the project, and in 2018 the Supreme Court ruled in their favour, stating that they should have been consulted prior to licenses for the project being granted. The community wants the project cancelled, though this looks unlikely given how far construction has advanced. Nonetheless, Grupo México has been forced to install much stronger safety measures than it had originally envisaged.

'The three relevant [environmental authorities] … say they have demanded higher protective barriers, the strengthening of these barriers and so on,' says Hopenhaym. 'So in theory, a series of measures has been taken, including measures to contain a possible disaster. We hope we won't ever have to see them put to the test.'

For when these measures are absent or inadequate, the consequences may be nothing short of catastrophic, as the people of Minas Gerais, Brazil, know only too well.

II. The Bento Rodrigues dam disaster, Brazil

'When they told us the dam had broken we knew something was heading our way, but we were expecting water,' says Marino D'Angelo Junior, a dairy farmer from the state of Minas Gerais, Brazil. 'What came was something from another world. A monster. I saw it.'

At 4:20 p.m., on 5 November 2015, a tailings dam failed at the Complexo do Germano iron ore mine, around 28 kilometres away from Paracatu de

Photo 5.2 Bento Rodrigues following the disaster / Romerito Pontes 2015 / CC BY 2.0

Cima, the village where D'Angelo was living with his wife Maria and daughter Alice. A tidal wave of thick brown slurry went flooding into the rolling hills and valleys of central Minas Gerais, destroying everything in its path. Bento Rodrigues, the village closest to the mining complex, was virtually wiped off the map. Nineteen people were killed and one heavily pregnant woman lost her baby.

Later that evening, a fire brigade helicopter landed on a football pitch in Paracatu de Baixo, just downhill from the D'Angelo family residence. An official descended and informed residents that they had just ten minutes to run for the higher ground (Carvalho, 2015).

'The mud arrived at around 7 p.m.,' explains D'Angelo. 'We were about 500 metres away. Then there was no need for any warning. There was a terrible smell of rotting oil, rotting mud, and a roaring sound … My father-in-law's house is just over there; everything was destroyed.'

Much of Paracatu de Baixo was destroyed, as were several other hamlets in the area. Around 500 families were made homeless in total.

The dam released 43.8 million cubic metres of tailings, which flowed into the Rio Gualaxo do Norte, then the Rio do Carmo, then finally the Rio Doce, the main river in the region. It travelled 650 kilometres along the entire length of the river, devastating villages, towns, and cities as it went. Seventeen days later, it reached the Atlantic Ocean at Regência, in the neighbouring state of Espírito Santo, with the tailings forming a plume which travelled out to sea and polluted the coastline both sides of the mouth of the river.

In total, the dam collapse affected 40 municipalities in the Rio Doce basin, an area roughly the size of Portugal. Globally, it is the largest tailings dam failure of all time (Robson, 2017, p. 30), and, as of early 2022, it remains Brazil's worst-ever environmental disaster.

'Everything is yet to be done'

The Complexo do Germano was run by Samarco, a joint venture belonging to the Brazilian company Vale and the Anglo-Australian BHP, both among the most powerful mining companies in the world. In the aftermath of the disaster, the three companies set up the Renova Foundation in an agreement with the Brazilian state. Renova's mission was to repair the damage caused and compensate those affected, including full resettlement of those who lost their homes.

But given that it was set up by the very same companies responsible for the disaster, Renova had a conflict of interest built into it from the start, something which has not been lost on the *atingidos* – the affected people – and their allies.

'Renova is a shield, which protects the companies in a formal, administrative and legal sense,' says Thiago Alves da Silva of the Movement of People Affected by Dams (MAB), which has done extensive work with the communities affected by the dam failure (Oliveira, 2020).

D'Angelo is more forthright: 'Everything that Renova does is to clean the blood from Samarco's reputation.'

Of course, true reparation for a disaster of this magnitude is impossible. The losses, hardships, and trauma that the *atingidos* have suffered go far beyond the material. The mud buried not only physical property such as houses and vehicles; it also destroyed communities, permanently altered relationships, terminated life projects. In this sense, much of the damage is irreparable.

Mônica dos Santos is a former resident of Bento Rodrigues, now studying to be a lawyer in light of her experience since the dam failure.

'They could build us mansions and pay us millions in compensation and it still wouldn't erase what we've been through, what we're still going through. It still wouldn't bring back the old life, the things we lost,' she says.

All the more reason that where the companies are able to mitigate the damage, they must do so fully, in a manner sensitive to the needs of those affected. But this hasn't happened. On the contrary, since the disaster, the *atingidos* have complained repeatedly – backed up by prosecutors at both state and federal level – that Renova has failed to do even the basics.

'Five years later, nothing is concluded; everything is yet to be done,' said Silmara Goulart, head of the Rio Doce taskforce at Brazil's Federal Prosecution Service (MPF), speaking at a press conference in October 2020.

'Not a single group of people affected — farmers, laundresses, artisans, fishers, small businesspeople — was fully compensated. Nor has the environment been fully recovered. Not even the municipality of Bento Rodrigues, symbol of the disaster, has been rebuilt.' (Tokarnia, 2020)

A litany of failures

Perhaps the most obvious of Renova's many failures is that there are still thousands of affected people who have not been registered and who still have no access to processes of compensation.

According to the consultants Ramboll, which was hired by the MPF to monitor and evaluate Renova's work, of the 83,838 applications for registration Renova had received by April 2021 (the most recent month for which figures are available), just 31,712 (38 per cent) had been accepted (Ramboll, 2021, p. 5). Of these, just 11,121 (35 per cent) had received some form of compensation (ibid., p. 7). This means that most of the applicants are still waiting for their cases to be processed.

Certain groups have struggled to obtain recognition as *atingidos* – such as fishing communities further downriver in Espírito Santo and along the coastline – even though it was clear they had been affected. Municipalities in Espírito Santo such as Serra and Conceição da Barra were not included in the initial agreement and residents had to fight for more than a year to be legally recognized (Haniff, 2021). Ramboll and the MPF also highlight the failure to register affected workers in several other informal sectors, including artisans, laundresses, and shellfish pickers, as well as some traditional communities which depended on the river for subsistence, such as indigenous people, *ribeirinhos* (riverside dwellers) and *quilombolas* (residents of Afro-Brazilian communities originally founded by escaped slaves) (Século Diário, 2019).

Not to be confused with the programme of compensation, the companies have been paying emergency maintenance to people in the Rio Doce basin whose livelihoods were affected by the disaster since late 2015. This is equivalent to the value of a monthly minimum wage (currently R$1,212, around $243), plus 20 per cent for each dependent, and the value of a basic monthly food package. However, many people in the Rio Doce basin have been unable to claim, because their work is informal and they are unable to provide official paperwork – and women are disproportionately affected.

Maria D'Angelo, Marino's wife, is a case in point. 'I always made lunch and dinner for Marino and his employees,' she says. 'If there were four of them on the job, I had to provide eight meals. That meant I was contributing, as my husband didn't have to buy food from elsewhere. But Renova didn't consider things like this.'

This problem has affected not only women in agricultural communities like Maria, but also those in fishing communities who carry out tasks such as cleaning fish and preparing nets. As a result, pre-existing gender inequalities in the Rio Doce basin have been exacerbated, with women increasingly dependent not only on their husbands, but even on their sons.

Many *atingidos* complain that the support on offer is derisory. According to Ramboll, the average monthly income of affected families in the Rio Doce basin has fallen by 49 per cent since the disaster. For low-income families, the situation is even worse, with the average fall in income at 73 per cent (Parreiras, 2020). Even *atingidos* like the D'Angelo family, who were relatively well off before the disaster, have found themselves struggling.

'I had an income of R$22,000 to R$23,000 per month [$5,720-$5,980],' says D'Angelo. 'Taking away my expenses – wages for my employees, feed for my animals – that left R$10,000 [$2,600] for me. Now look what I've been given this month – R$1,115 [$290]. And I still have to pay my employee.'[2]

This hasn't stopped Renova from attempting to cancel or reduce the emergency assistance on at least three occasions (Viana and Cobuci, 2021). For example, in July 2020, at the height of the first wave of the Covid-19 pandemic, Renova unilaterally cancelled the aid to 7,000 subsistence fishers and farmers, arguing that conditions had been re-established for them to return to their former activities (Pimentel, 2020). The courts disagreed and forced Renova to reinstate the payments, though since January 2021, the value has been cut by half for these groups.

But just like Grupo México, Renova has devoted considerable resources to public relations. In May 2021, state and federal prosecutors and public defenders highlighted that Renova had spent R$28 million ($5.3 million) on its communications strategy – including R$17.4 million ($3.3 million) on a single contract in 2020 – with the aim of promoting a positive image of Renova and its parent companies across the Brazilian media. Demanding that this propaganda be withdrawn, the prosecutors and public defenders argued it presented information that was 'imprecise, unreliable, incomplete or mistaken' and it 'testified to the re-establishment of a non-existent normality' (Rodrigues, 2021a).

Waiting in vain

'Bento was a sub-district of Mariana, home to around 600 residents. We had a church dating back to the 18th century, from where our patron saint São Bento was lost in the disaster,' says Mônica dos Santos. 'We were a very united community ... Now, everyone's been scattered.'

Like most of those who were made homeless by the disaster, Dos Santos moved into temporary accommodation in Mariana, the nearest town of any significant size. Over six years later, she is still there. In February 2021, Renova missed the deadline for completion of the reconstruction of Bento Rodrigues and Paracatu de Baixo for the third time (Nascimento, 2021). In October 2022, Renova announced that the infrastructure in Bento Rodrigues was complete and invited the *atingidos* to occupy the town from January 2023 (Fundação Renova, 2022; Oliveira, 2022). However, with only 71 of the 162 houses having been finished, the *atingidos* complain they have been invited to live on a building site (Oliveira, 2022). Paracatu de Baixo is even further behind, with not a single house completed (Mansur and Franco, 2022).

Renova cited Covid-19 as an excuse for failing to meet its 2021 deadline, but it had been obvious since long before the pandemic that the work was not going to be ready on time. In October 2017, I met with MAB's Thiago Alves da Silva, who was in London to demand some answers at BHP's AGM.

'The resettlements are a long way behind schedule,' he said. 'The companies gave themselves a deadline of 2019 to finish the work, but they didn't say when they would start it. That's true of Bento Rodrigues, Paracatu de Baixo and Gesteira, a tiny hamlet of nine houses, a church, and a community centre. It's a single street. It's unthinkable that BHP Billiton, which put on this event today, can't build nine houses.'

According to a report by Renova in March 2021, out of 557 families it had identified for resettlement, it had completed the process with just 15 (Ramboll, 2021, p. 21). None of the former residents of Bento Rodrigues and Paracatu de Baixo have yet been able to return home.

'The people from these communities moved to Mariana, a city of 70,000 inhabitants,' said Da Silva. 'This is good for the company, because it leads the families to give up on resettlement. They lose hope, start to integrate in the city, and end up asking for financial compensation rather than resettlement. This is much cheaper for the company and means they avoid long-term compensation processes.'

Delay is not accidental; it is a deliberate strategy, with the aim of wearing down the *atingidos* so they accept whatever is on offer. Letícia Aleixo is a lawyer who works with the Catholic NGO Cáritas. She has seen this strategy play out first-hand.

'Everyone in Mariana is depressed,' she says. 'People just want their homes back. They've all fallen out with each other. People have no strength left to refuse any money; they want whatever handout they can get. And they even think the company's doing them a favour! They're exhausted.'

Divide and rule

With so many different communities affected by the disaster, over such a vast area, there was an urgent need to establish consistent and transparent systems of registration and compensation. But again, this hasn't happened. Ramboll has criticized Renova for 'lack of transparency in the process of registration, which has generated insecurity, rumours and conflicts within the communities' (Século Diário, 2019).

'I've lived in Mariana for two years, going to all these meetings, and the different commissions [of *atingidos*] fight amongst themselves,' says Aleixo. 'We've even had to physically separate people, because the company will make an offer to one group and offer something different to another.'

What has helped to dispel the confusion created by Renova is the provision for independent technical advisers – such as Cáritas, which has been providing support to the *atingidos* in Mariana.

The technical advisers are professionals with a degree in a relevant field, chosen by the *atingidos* 'to ensure broad and well-informed participation in decision-making processes and to support affected communities in the struggle for full compensation for losses and damages suffered' (Cáritas Minas Gerais, 2019, p. 24). The courts obliged Renova to provide the funding, though the advisers are independent. The aim is to equip the *atingidos* with good information, making them understand what their rights are and helping them work out how they might re-establish dignified and self-sufficient living conditions.

'I was just a rural worker, a housewife,' says Maria D'Angelo. 'I had my life on my land, which was totally ruined by the mine tailings, and I was unable to argue with the companies the way I do now. Without the advice of qualified people, we would be destitute today' (ibid., p. 25).

But Renova has resisted the implementation of this system (Moreira et al., 2020, p. 10). When it was created, the Rio Doce basin was divided into 21 areas. Currently, just three are receiving this support. Without access to technical advice, the *atingidos* find themselves unsure of what to believe and under pressure to accept whatever Renova is offering. These offers are usually conditional upon the signing of release clauses, meaning Renova is freed of any future obligations and the values involved cannot be revised – values which are typically very low when set against the loss of homes, property, and livelihoods.

'The fishermen who fished just to eat, often without selling much, received R$11,200 [$2,900],' says the activist Letícia Oliveira, also of MAB. 'And that was it, after that they were on their own. The company did nothing to check if they'd be able to fish again, sell the fish or even just eat it.'

In late 2020, with so many *atingidos* still waiting to be registered, the courts established a new, simplified system of compensation, to facilitate access for informal workers who have difficulties proving their losses. In little more than ten months, it had doled out compensation to nearly 50,000 people (Centro Alternativo de Formação Popular Rosa Fortini, 2021). While this money has been a lifeline for a local economy badly hit by the disaster and then the pandemic, critics argue the system is abusive.

For the prosecutor André Sperling, these values are 'ridiculously low, which in no way constitute full reparation' (Maciel, 2021). The Federal Prosecution Service (MPF) attempted to intervene, arguing that the compensation granted to each *atingido* via the new system should be considered only a minimum. A federal judge disagreed, however, and the new system remains in place.

In February 2021 state prosecutors from Minas Gerais filed a request for the abolition of the Renova Foundation. Echoing MAB's assessment, the document argues that 'the Foundation has acted far more as a means of limiting the responsibilities of its parent companies (Vale and BHP Billiton) than as an agent of effective human, social and environmental reparation' (UOL, 2021).

'The crime is renewed'

Despite the hard work of prosecutors at both state and federal level to defend the rights of the *atingidos*, the Brazilian state has shown systematic unwillingness to hold Samarco, Vale, and BHP to account.

Not only has Samarco managed to dodge most of its liabilities towards the *atingidos*, the company has not even paid the fines it was hit with in the aftermath of the disaster. As of December 2019, Samarco had paid just R$72 million ($17.3 million) out of the R$301 million ($72.2 million) in fines it owed to the Brazilian state, with the company contesting R$180 million ($43.2 million) in the courts (Oliveira, 2019).

In 2016, almost a year after the disaster, federal prosecutors charged 21 executives from Samarco, Vale, and BHP with homicide and bodily harm. The accused included Ricardo Vescovi, Samarco president at the time of the disaster, and Kleber Terra, the company's head of operations.

But these charges were eventually thrown out in April 2019, though prosecutors insisted there was enough evidence to proceed with the case. Samarco's internal communications show not only that the company knew the dam was a risk, but even reveal a grimly precise awareness of what the consequences of a failure would be: one report predicted up to 20 deaths (Rodrigues, 2019a). Charges have been dropped against 16 of the original 21 defendants, and those who remain will now respond only for lesser, environmental crimes. To this day, no-one has ever spent any time in jail.

Yet more salt was rubbed in the open wounds of the *atingidos* when Samarco was granted permission to restart operations in late 2019, a decision which was made public just days before the fourth anniversary of the disaster. By December 2020, the company was up and running again. Little wonder that in November 2020, on the fifth anniversary of the disaster, Cáritas put out a press release entitled 'O crime se renova' (literally, 'the crime is renewed'), highlighting the ongoing hardships faced by the *atingidos* and denouncing the Renova Foundation for perpetuating them.

Given the total failure to obtain justice in Brazil, more than 200,000 *atingidos* from the Rio Doce basin have decided to pursue BHP in the UK courts, in one of the largest class action cases in English legal history. Though the High Court initially denied jurisdiction in the UK, in July 2022 the Court of Appeal overturned the ruling, meaning that the *atingidos* will have their day in an English court. They are demanding £5 billion in compensation.

'We're still facing consequences on a daily basis,' says Marino D'Angelo. 'The crime wasn't only on the 5 November [2015]; it's brought consequences for us every day of our lives. And as time passes, they get worse.'

III. The Brumadinho dam disaster, Brazil

To the untrained eye, the dam looks like nothing more than a grassy patch on a hillside, roughly pentagonal in shape. While it does stand out from the tree-covered hills that surround it, it could just be land deforested for timber, or for pasture.

Then it collapses.

It takes the brain a moment to register what it is seeing: that this grassy patch wasn't solid earth at all, but liquid.

As the whole structure folds, like a building being demolished, a tsunami of brown slurry goes tearing down towards the mining site at the bottom of the shot, consuming everything in its path. The forest below the dam turns from green to brown as the tailings descend, while at the top of the image, a gaping pit in the hillside becomes more visible by the second.

Another video, taken from a crane overhanging the mining site, shows vehicles attempting to flee the scene as the tailings come flooding in. It looks nightmarish, unreal, like people trying to flee a volcanic eruption in a scene from a disaster movie.

These images, which are widely available online, were filmed not at the Complexo do Germano, but at another iron ore project located 66 km to the west, at Córrego do Feijão, near the town of Brumadinho, where a second tailings dam failed on 25 January 2019.

Initially, history seemed to be repeating itself: here was another catastrophic tailings dam failure, also at an iron-ore operation, also in Minas Gerais, and also with the involvement of Vale – in this case the sole owner and operator of the mining complex.

But the death toll of this second disaster was much higher.

Most of the victims were employees at Córrego do Feijão. It was lunchtime when the dam failed, and the workers' canteen stood directly in the path of the tailings. Several nearby homes and a guesthouse were also buried. In total, 270 people were killed, including two women who were expecting children.[3]

Nearly four years later, not all of the bodies have been recovered. In December 2022, police and firefighters identified the 267th victim. They continue to sift through the tailings in search of the three who are still missing.

'The siren never sounded'

'When before there was peace in the town and familiarity amongst people, a place where almost everyone knows each other, what came to dominate was pain,' says Marcela Rodrigues. Originally from the town of João Monlevade, she moved to Brumadinho with her family as a child.

'That's what we see today in Brumadinho. Even if you haven't lost anyone, everyone knows of someone who died – a relative of one of your friends, perhaps just someone you'd say hello to at the bakery.'

One of those who died was Rodrigues' father, Denilson. She slid a photo across the table to me, when I met her in London in late 2019. In the photo Denilson was smiling, his arms folded across his chest. He looked relaxed, happy, much younger than his 49 years. A Vale employee, he was in the canteen at the time of the collapse.

'He'd gone into work to do safety training that same day,' Rodrigues sighed. 'Nearly a month later we found two pieces of his body.'

She is highly critical of the way she was treated by Vale in the aftermath of the disaster.

'The company only offered me eight sessions with a therapist. Even then I had to fight for the right to have them with my own therapist,' she said. 'They wanted to assign me a psychologist – this happened to my brother – who turned up at the house and started to justify the dam collapse. He said that we could have lost our dad in any other set of circumstances. So it's all a lie, they don't really care about people at all, and they have no sense of responsibility towards them.'

The impacts on mental health for those affected by both tailings dam failures have been severe, with *atingidos* widely reporting symptoms of anxiety and depression. In Brumadinho, where so many lost friends or family members, the situation is particularly acute. Survivors, witnesses, and bereaved family members have been suffering from PTSD symptoms similar to those experienced by survivors of war (*Estado de Minas*, 2019). According to the municipal health secretariat, the use of antidepressants increased by 56 per cent and anxiolytics by 79 per cent in 2019, compared to the previous year. Suicide attempts rose from 29 to 47 (Freitas and Almeida, 2020).

Andrew Hickman is a trustee at London Mining Network who visited Brumadinho in the wake of the disaster. 'Having been in Brumadinho for three or four days, it was clear that everyone – the whole town – even those who weren't directly affected, were completely in trauma,' he says. 'People are so shocked, and so traumatized, they are just trying to get by one day at a time, to get through the next conversation.'

Vagner Diniz, an IT manager from São Paulo, also lost loved ones in the disaster. His son Luiz, daughter Camila and daughter-in-law Fernanda had come to Brumadinho to visit Inhotim, a well-known open-air art gallery and botanic garden just outside town. They were staying at the guesthouse that was buried by the tailings.

'But the siren never sounded,' Diniz wrote on Facebook in the aftermath. 'The manager of the guesthouse told us that they were very well prepared for an emergency and would use all the escape routes. In five minutes everyone would be safe. But the siren never sounded.'

Luiz and Fernanda were expecting their first child, Lorenzo. The image accompanying Diniz's post shows a smiling Luiz holding up a pair of baby shorts.

Like Rodrigues, Diniz is scathing of the way he and his wife Helena were treated by Vale. 'Vale never showed any interest in talking to us,' he says.

'They never contacted us. Four months later we decided to take them to court. My children were not Vale workers. They were not an outsourced company's workers. They didn't belong to a community in Brumadinho affected by the mud. We thought, "We are forgotten." And so we went to court.'

Diniz and his family demanded not only financial reparations from the company, but also what he calls 'moral reparation'. 'We asked for Vale to recognize that they were guilty, they committed a crime,' he explains. 'We demanded that Vale build a memorial in every Vale office around the world with a picture of the people who died. And we also demanded a minute's silence before the beginning of every Vale AGM. To ask for moral reparations, for us, is a way of Vale never forgetting what happened.'

A judge in Minas Gerais obliged Vale to pay Diniz and his family compensation – albeit just 20 per cent of the sum requested – but rejected their request for moral reparations, on the grounds that such a demand should come from the *atingidos* as a collective (Gatehouse, 2020).

A flood of money

In the wake of the disaster, the courts ordered Vale to pay emergency relief to all residents of Brumadinho without distinction. This was the value of a monthly minimum wage for each adult, half this total for each adolescent, and a quarter for each child. In the other affected municipalities, the relief was provided to all residents living within a kilometre of the Rio Paraopeba, the river flooded by the tailings (Rodrigues, 2021b).

In total, nearly 108,000 people were offered this assistance, which was guaranteed initially for a year. In November 2019, pressure from the *atingidos* helped to get it extended, though the values were reduced by half for most of the recipients – between 93,000 and 98,000 people – as they were considered not to have been directly affected (Rodrigues, 2019b).

While the aid has been a lifeline for those who lost their livelihoods due to the disaster, it has also created problems.

'There are many who want justice, who fight for justice. They want from Vale more effective measures to rehabilitate the city. They don't want this kind of pocket money every month,' says Diniz. 'There is a division in the city: those who are living this euphoria, because there's a lot of money flowing, and those who want justice.'

Vale has also used this money as a bargaining chip, as a means of resisting a proper, independent assessment of the damage and of getting entangled in long-term processes of compensation and reparation. For example, while the right of affected communities to specialist technical advice – as in the case of the 2015 disaster – was established in a court agreement in February 2019, Vale spent the following year dragging its feet, arguing for a reduction of the groups' budgets, their remits, and the period of time in which they would be allowed to operate (Weimann, 2020).

'In the court hearings for Brumadinho, what they're arguing is that technical advice for the communities is unnecessary because Vale has already paid 50 per cent of the compensation it owes.' says Letícia Aleixo of Cáritas. 'What's going on here? They're trying to define who the affected are. But who can say yet who has and hasn't been affected? We know which families lost people in the disaster. But we don't know who's unable to work, who has no water, who's become ill. There's a lot of damage. They want to have control: who's to be compensated, how much, and when.'

While the groups of technical advisers are now active in the Rio Paraopeba basin, they are working according to very specific remits. The pandemic has also made matters more difficult. Throughout most of 2020 and into 2021, much of the requisite fieldwork was postponed and meetings between *atingidos* and their advisers had to be held online – excluding those who had no internet access.

The atingidos *lose out*

It wasn't until well into 2020 that most of the *atingidos* in Brumadinho and other communities in the Rio Paraopeba basin even began the long process of assessing the damage and determining how it might be repaired. And yet, in February 2021, Vale signed an agreement with the state of Minas Gerais for reparations for the Brumadinho disaster. It is worth R$37.7 billion ($7.2 billion) – the largest agreement of its kind in Latin American history.

'We've managed to finalize the largest agreement in Brazilian history in terms of establishing compensation and socioenvironmental reparation,' said Gilson Lemes, president of the Minas Gerais Court of Justice. 'The conflict was resolved through mediation and conciliation, neutrally and impartially, with dialogue between the parties concerned' (TJMG, 2021).

But for the *atingidos*, the conflict is far from resolved, and nor were they even one of the parties invited to the table for dialogue.

'After a crime which killed 272 people and destroyed the way of life of thousands more, leaving communities without water and in a deep state of trauma, it's outrageous that an agreement can be signed without the participation of those affected,' says MAB's Joceli Andrioli (MAB, 2021).

Vale's payment is considerably less than the sum the government had initially sought, which was around R$54 billion ($10.3 billion) (Lolli, 2021). Vale was also permitted to subtract the value of payments it had made under the emergency relief scheme, though for the *atingidos*, this assistance was unconditional and should not have been on the table during negotiations (Dotta, 2021).

Around 30 per cent of the R$37.7 billion will be used to benefit the population of Brumadinho, including a cash transfer programme to replace the system of emergency relief. This came into effect in November 2021 and will attend close to 100,000 people (G1, 2021). But the government of Minas Gerais is also planning to spend R$3.75 billion of the money on an

Photo 5.3 Indigenous *atingidos* assess the damage to the Rio Paraopeba / Felipe Werneck/ IBAMA 2019 / CC BY-SA 2.0

orbital highway around the state capital Belo Horizonte, as well as another $1.2 billion on improving mobility within the city, including expanding the urban metro system (Girundi et al., 2021). Another R$4.3 billion will be spent on public services which have no direct connection to Brumadinho or the Rio Paraopeba basin (Dotta, 2021).

For the *atingidos* and their allies, this is a betrayal. MAB swiftly filed an appeal with the Supreme Court alongside other social movements, with the support of the left-wing Workers' Party (PT) and Socialism and Liberty Party (PSOL). However, the appeal was rejected.

'With these funds, the governor is pushing ahead with a number of public works in the state, with his eye on the next election campaign,' said Marcelo Barbosa, of the Movement for Popular Sovereignty in Mining and a resident of Brumadinho. 'It's the affected communities who lose out' (Dotta, 2021).

'It will happen again'

In 2020, the courts agreed to hear a case brought by prosecutors for Minas Gerais against Vale and Tüv Süd, a German auditing company which signed a safety declaration for the dam at Córrego do Feijão just four months before it collapsed. According to the Brazilian Federal Police, Tüv Süd knew that the state of the dam was incompatible with good engineering practice and did not comply with Brazilian safety regulations (Rodrigues, 2021c).

A report by the Federal Police and an independent study carried out by the Universitat Politécnica de Catalunya in Spain also highlight the role of a third company, the Dutch multinational Fugro, which was contracted by Vale for a borehole drilling operation on the dam. Both studies conclude that a borehole drilled on the day of the collapse was the trigger for the failure. According to the Federal Police report, the companies knew the procedure was risky but went ahead with it anyway (ibid.).

Sixteen employees of Vale and Tüv Süd were charged with homicide, including Fabio Schvartsman, Vale CEO at the time of the disaster, and Chris-Peter Meier, general manager of Tüv Süd. Based in Germany, Meier has refused to travel to Brazil to provide a statement to police (Rodrigues, 2021d). As of December 2022, these charges still stand, though the process is proving painfully slow.

Even if the criminal cases do go ahead, the *atingidos* and their allies insist that true justice would demand genuine structural change, with the reasons for the two disasters going far beyond a simple question of poor management.

'It would be symbolic, if someone high ranking went to jail,' says Letícia Aleixo. 'But that doesn't solve things. People have the feeling that someone has been punished, but if it hadn't been that particular president in charge it would have been someone else and things would've unfolded in exactly the same way.'

An analysis by *The New York Times* after the disaster revealed that there were 87 dams in Brazil built using the same method as the dams that collapsed in Bento Rodrigues and Brumadinho. Known as upstream tailings construction, it is the cheapest method of tailings containment, in which the retaining structures are built upon the tailings themselves. In the case of at least 27 of these dams, there are settlements of more than 1,000 people located downhill and within eight kilometres. This means that the lives of as many as 100,000 people could be at risk (Bloch et al., 2019).

In 2020 Congress did finally pass a new law on dam safety. Though it was watered down to an extent, it is an improvement on what came before. The construction of new upstream tailings dams was banned, and companies were given until February 2022 to decommission existing dams built using this method. However, this is leading to displacement, as it is easier for companies to relocate communities than to repair or decommission dangerous dams (Christian Aid, 2021, p. 5).

By the end of 2021, Brazil had 40 dams which were considered to be in a state of emergency. Of these 40, three were ranked Level 3, the most critical – in other words, where the dam has already ruptured or is at imminent risk of doing so. All three are in Minas Gerais, and all three belong to Vale (*ISTOÉ Dinheiro*, 2022).

'With everyone we've talked to, we've been very clear: it will happen again,' says Mônica dos Santos. 'Bento wasn't the first and it won't be the last. And Brumadinho won't be the last.'

***

Mining is an inherently dangerous business. Extraction depends on the widespread use of heavy machinery and explosives, which erode the soil and pollute water sources with sedimentation. There is a risk of acid mine drainage at mine sites and at tailings storage facilities. Processing of ore involves the use of chemicals such as cyanide, mercury, and sulphuric acid, all of which are highly toxic to humans and to wildlife. Smelters release toxins such as sulphur dioxide and carbon monoxide, as well as heavy metals which may bio-accumulate in the atmosphere. And tailings storage facilities require constant monitoring and control long after a mine has ceased operations.

But for far too long the mining industry has been allowed to regulate itself. Even where robust environmental regulation exists, all too often there is a failure of enforcement, with governments allocating insufficient resources to monitoring and oversight of mining operations. Mexican authorities hadn't inspected the installations at Buenavista de Cobre for at least 14 years prior the 2014 spill, for example (Cárdenas, n.d.,a). As is clear from these recent cases, there is a chronic lack of political will when it comes to regulating the industry and holding companies to account when disasters occur.

Since the disaster at Brumadinho, there has been an increased concern amongst investors regarding tailings management, with greater pressure on the industry to adopt tougher safety standards and provide detailed disclosure on tailings storage facilities. While this is welcome and long overdue, much more must be done to reform a deeply entrenched economic model which prioritizes investment and business growth over the human rights of communities in mining regions. For now, there is no sign of a shift away from the failed system of self-regulation that has had such disastrous consequences for the communities covered here in Sonora and Minas Gerais.

But the need for reform has never been more urgent. With the mining industry likely to expand significantly in the coming years due to rising demand for metals, an ever-increasing amount of waste will be generated, requiring the construction of even larger tailings dams, which are more likely to fail and more likely to cause major damage when they do. Climate change is also a factor, with extreme and unpredictable weather events also likely to increase the risk of dam failures. Severe flooding in Minas Gerais in January 2022 put locals on tenterhooks, highlighting once again the danger posed by tailings dams throughout the state.

But it must be stressed that these disasters usually occur not for isolated or exceptional reasons, but because companies cut corners: they ignore state regulation, disregard good engineering practices, and have only weak systems of oversight in place (Christian Aid, 2021, p. 2). There is little incentive for companies to reform, given the vast profits the industry continues to generate and the culture of near total impunity that prevails when things do go wrong. Nothing will change while such recklessness is rewarded.

Notes

1. While boiling water kills viruses and bacteria, it does nothing to eliminate heavy metals.
2. The exchange rate used is that at the time of the interview, on 23/11/2018. Likewise, where other conversions are given, the exchange rate used is that of the time in question.
3. For this reason, the death toll is sometimes reported as 272, e.g., by MAB's Joceli Andrioli on p. 110.

References

All references to web-based material were checked and still available in November 2022 unless otherwise stated.

All references are listed, with clickable links for your convenience, on the page for this chapter on the Heart of Our Earth website: <https://lab.org.uk/the-heart-of-our-earth/>

Bloch, M., Reinhard, S. and Peçanha, S. (2019) 'Where Brazilians Live in High-Risk Areas Downhill From Mining Dams'. [online] *The New York Times*. Available at: <https://www.nytimes.com/interactive/2019/02/14/world/americas/brumadinho-brazil-dam-collapse.html>.

Blust, K. (2019) 'The Impact Of Mexico's Worst Mining Disaster, 5 Years Later'. [online] Fronteras. Available at: <https://fronterasdesk.org/content/1097126/impact-mexicos-worst-mining-disaster-5-years-later>.

Cárdenas, P. (n.d.,a) 'Informe Río Sonora: la omisión que quitó vida a miles'. [online] CONNECTAS. Available at: <https://www.connectas.org/especiales/rio-sonora/>.

Cárdenas, P. (n.d.,b) 'La peor tragedia de la industria minera en México'. [online] Fundar. Available at: <http://riosonora.fundar.org.mx>.

Cáritas Minas Gerais (2019) *Cáritas Minas Gerais in Common House Care. Facing the mining model and defending the social and environmental rights.* Belo Horizonte, MG.

Carvalho, P. (2015) 'O momento da tragédia'. [online] *Época Negócios*. Available at: <https://epocanegocios.globo.com/Brasil/noticia/2015/12/o-momento-da-tragedia.html>.

Centro Alternativo de Formação Popular Rosa Fortini (2021) 'Justiça Federal abre o Sistema Indenizatório Simplificado para todas as cidades atingidas e prorroga prazo de fechamento'. [online] Available at: <https://www.centrorosafortini.com.br/noticia.php?id=390>.

Christian Aid (2021) *The true cost of mining. Ensuring justice for people and communities affected by the Brumadinho dam disaster.* [online] Christian Aid. Available at: <https://www.christianaid.org.uk/sites/default/files/2021-01/Brumadinho%20Report.pdf>.

Comisión Federal para la Protección contra Riesgos Sanitarios (COFEPRIS) (2014) 'La Secretaria de Salud anuncia el inicio de la segunda fase de atención médica especializada para los afectados por el derrame en Río Sonora'. [online] Available at: <https://www.gob.mx/cofepris/prensa/la-secretaria-de-salud-anuncia-el-inicio-de-la-segunda-fase-de-atencion-medica-especializada-para-los-afectados-por-el-derrame-en-rio-sonora>.

Comités de Cuenca Río Sonora and PODER (2018) *Derrame de 40 millones de litros de solución de cobre acidulado a los ríos Bacánuchi y Sonora provenientes de las instalaciones de la mina Buenavista del Cobre, S.A. de C.V., subsidiaria de Grupo México.* [online] PODER. Available at: <https://poderlatam.org/wp-content/uploads/2018/05/Sonora_Informe_May_2018.pdf>.

Comités de Cuenca Río Sonora and PODER (2020) *Cronología de la Impunidad.* [online] Río Sonora and Mexico City: PODER. Available at: <https://poderlatam.org/wp-content/uploads/2020/08/Cronologia_delaImpunidad.pdf>.

Comités de Cuenca Río Sonora and PODER (2021) *Ante el incumplimiento del Fideicomiso Río Sonora exigimos #RemediaciónYARíoSonora.* [online] Río Sonora and Mexico City: PODER. Available at: <https://poderlatam.org/wp-content/uploads/2021/05/prensa_RPI_FRS_170521_fin.pdf>.

Contreras, A. (2020) 'La Corte ordena a Grupo México reabrir fideicomiso por daños en Río Sonora'. [online] Pie de Página. Available at: <https://piedepagina.mx/la-corte-ordena-a-grupo-mexico-reabrir-fideicomiso-por-danos-en-rio-sonora/>.

Díaz-Caravantes, R. E., Durazo-Gálvez, F. M., Moreno Vázquez, J. L., Duarte Tagles, H. and Pineda Pablos, N. (2021) 'Las plantas potabilizadoras en el río Sonora: una revisión de la recuperación del desastre', *Región y Sociedad*, 33, e1416. Available at: <https://regionysociedad.colson.edu.mx:8086/index.php/rys/article/view/1416/1746#toc>.

Dotta, R. (2021) 'Atingidos recorrerão ao STF para questionar acordo entre Vale e Governo de Minas'. [online] *Brasil de Fato*. Available at: <https://www.brasildefatomg.com.br/2021/02/04/atingidos-recorrerao-ao-stf-para-questionar-acordo-entre-vale-e-governo-de-minas>.

Estado de Minas (2019) 'Após lama, Brumadinho registra alta de suicídio e prescrição de remédios'. [online] Available at: <https://www.em.com.br/app/noticia/gerais/2019/09/09/interna_gerais,1083678/apos-lama-brumadinho-registra-alta-de-suicidio-e-uso-de-remedios.shtml>.

Freitas, R. and Almeida, F. (2020) 'Brumadinho convive com adoecimento mental um ano após tragédia da Vale'. [online] G1. Available at: <https://g1.globo.com/mg/minas-gerais/noticia/2020/01/21/brumadinho-convive-com-adoecimento-mental-um-ano-apos-tragedia-da-vale.ghtml>.

Fundação Renova (2022) 'BENTO RODRIGUES'. [online] Fundação Renova. Available at: <https://web.archive.org/web/20220926203006/https://www.fundacaorenova.org/reassentamentos/bento-rodrigues/>.

Fundar (2018) *Fideicomisos en México. El arte de desaparecer dinero público.* [online] Mexico City: Fundar. Available at: <https://fundar.org.mx/mexico/pdf/FideicomisosEnMexico.pdf>.

Fundar México (2018) *Fideicomiso Río Sonora – Fideicomisos.* [video] Available at: <https://www.youtube.com/watch?v=HzowgJjB8_4>.

G1 (2021) *Brumadinho: Atingidos deixam de receber auxílio e passam para programa de transferência de renda.* [video] Available at: <https://g1.globo.com/mg/minas-gerais/mg1/video/atingidos-de-brumadinho-ingressam-em-programa-de-transferencia-de-renda-10009916.ghtml>.

Gatehouse, T. (2020) 'Brumadinho – one year on'. [online] Latin America Bureau. Available at: <https://lab.org.uk/brumadinho-one-year-on/>.

Girundi, D., Freitas, R. and Moreno de Castro, C. (2021) 'Vale assina acordo de R$ 37,68 bilhões para reparar tragédia de Brumadinho'. [online] G1. Available at: <https://g1.globo.com/mg/minas-gerais/noticia/2021/02/04/

vale-assina-acordo-bilionario-de-r-3768-bilhoes-para-reparar-danos-causados-em-brumadinho.ghtml>.

Haniff, J. (2021) 'Water for life, not for death'. [online] Latin America Bureau. Available at: <https://lab.org.uk/water-for-life-not-for-death/>.

ISTOÉ Dinheiro (2022) 'País encerra 2021 com 40 barragens em situação de emergência'. [online] Available at: <https://www.istoedinheiro.com.br/pais-encerra-2021-com-40-barragens-em-situacao-de-emergencia-declarada-diz-anm/>.

Lamberti, M. (2018) *Análisis del Fideicomiso Río Sonora. Simulando la remediación privada en un Estado capturado*. [online] Mexico City: PODER. Available at: <https://poderlatam.org/wp-content/uploads/2020/01/analisis-FRS.pdf>.

Lolli, J.F. (2021) 'Acordo entre Vale e governo recebe duras críticas de atingidos por barragem de Brumadinho'. [online] itatiaia. Available at: <https://www.itatiaia.com.br/noticia/acordo-entre-vale-e-governo-recebe-duras-criticas-de-atingidos-por-barragem-de-brumadinho>.

Maciel, A. (2021) 'Áudio revela ameaças e intimidação de advogada da Renova aos atingidos pelo desastre de Mariana'. [online] Agência Pública. Available at: <https://apublica.org/2021/02/audio-revela-ameacas-e-intimidacao-de-advogada-da-renova-aos-atingidos-pelo-desastre-de-mariana/>.

Mansur, R. and Franco, L. (2022) 'Mariana: atingidos por tragédia fazem protesto durante anúncio de entrega parcial de casas; 'não cabe uma cama de casal''. [online] G1. Available at: <https://g1.globo.com/mg/minas-gerais/noticia/2022/10/19/mariana-atingidos-por-tragedia-fazem-protesto-durante-anuncio-de-entrega-parcial-de-casas-nao-cabe-uma-cama-de-casal.ghtml>.

Méndez, E. and Sánchez Dórame, D. (2018) 'Tinacos, a sobreprecio; derrame en el Río Sonora 2014'. [online] *Excélsior*. Available at: <https://www.excelsior.com.mx/nacional/tinacos-a-sobreprecio-derrame-en-el-rio-sonora-2014/1257265>.

Moreira, R., Momm, E. and Leitão, K.O. (2020) 'O papel e limite das assessorias técnicas independentes no desastre-crime da Samarco (Vale/BHP Billiton)'. [online] 44º Encontro Anual da ANPOCS. Available at: <https://www.anpocs2020.sinteseeventos.com.br/arquivo/download public?q=YToyOntzOjY6InBhcmFtcyI7czozNToiYToxOntzOjEwOiJJR-F9BUlFVSVZPIjtzOjQ6IjQ5MTAiO30iO3M6MToiaCI7czozMjoiNDQ1NT QxMjU4YzE3NWNhZmIxMDhiZjIzNDA1OWRhYzki O30%3D>.

Movimento dos Atingidos por Barragens (MAB) (2021) 'Prejudicados por acordo bilionário, atingidos pelo crime da Vale em Brumadinho acionam STF'. [online] Available at: <https://mab.org.br/2021/02/10/prejudicados-por-acordo-bilionario-atingidos-pelo-crime-da-vale-em-brumadinho-acionam-stf/>.

Nascimento, P. (2021) 'Novo Bento continua no papel, e distrito é "cidade-fantasma"'. [online] *O Tempo*. Available at: <https://www.otempo.com.br/cidades/novo-bento-continua-no-papel-e-distrito-e-cidade-fantasma-1.2565437#>.

Oliveira, C. (2019) 'De R$ 301 milhões em multas aplicadas à Samarco, R$ 229 milhões ainda não foram pagos'. [online] *Hoje em Dia*. Available at: <https://hojeemdia.com.br/horizontes/de-r-301-milhões-em-multas-aplicadas-à-samarco-r-229-milhões-ainda-não-foram-pagos-1.761587>.

Oliveira, R. (2022) 'Sobreviventes de Bento Rodrigues são convidados a viver em um canteiro de obras'. [online] *O Tempo*. Available at: <https://www.otempo.com.br/cidades/sobreviventes-de-bento-rodrigues-sao-convidados-a-viver-em-um-canteiro-de-obras-1.2752496>.

Oliveira, W. (2020) 'Em cinco anos, história do crime de Mariana (MG) é marcada pela injustiça'. [online] *Brasil de Fato*. Available at: <https://www.brasildefato.com.br/2020/10/30/em-cinco-anos-historia-do-crime-de-mariana-mg-e-marcada-pela-injustica>.

Parreiras, M. (2020) 'Justiça inglesa frustra expectativas de atingidos da tragédia de Mariana'. [online] *Estado de Minas*. Available at: <https://www.em.com.br/app/noticia/gerais/2020/11/09/interna_gerais,1202753/justica-inglesa-frustra-expectativas-de-atingidos-da-tragedia-de-maria.shtml>.

Pimentel, T. (2020) 'Renova suspende auxílio a 7 mil atingidos pela tragédia de Mariana em MG e no ES; MPF abre inquérito'. [online] G1. Available at: <https://g1.globo.com/mg/minas-gerais/noticia/2020/07/08/defensorias-publicas-de-mg-e-es-alem-do-mpf-pedem-explicacoes-a-renova-que-suspendeu-auxilio-a-7-mil-atingidos-pela-tragedia-de-mariana.ghtml>.

PODER (2015) *Informe de Investigación. Análisis de incumplimientos de la normatividad ambiental por parte de Buenavista de Cobre, S.A. de C.V.*. [online] Mexico City/New York: PODER. Available at: <https://poderlatam.org/wp-content/uploads/2015/09/Análisis-de-incumplimientos-de-la-normatividad-ambiental-por-parte-de-BDC.pdf>.

PODER (2020) 'Cofepris reconoce contaminación masiva en agua de consumo humano en el Río Sonora'. [online] Available at: <https://poderlatam.org/2020/11/cofepris-reconoce-contaminacion-masiva-en-agua-de-consumo-humano-en-el-rio-sonora/>.

Ramboll (2021) *Relatório de Monitoramento Mensal – Mês 052 – Abril/2021. RELATÓRIO DE MONITORAMENTO MENSAL DOS PROGRAMAS SOCIO ECONÔMICOS E SOCIOAMBIENTAIS PARA REPARAÇÃO INTEGRAL DA BACIA DO RIO DOCE.* [online] Available at: <http://www.mpf.mp.br/grandes-casos/caso-samarco/documentos/relatorios-ramboll/pr-mg-00023462_2021.pdf>.

Robson, P. (2017). *The river is dead. The impact of the catastrophic failure of the Fundão tailings dam.* [online] London: London Mining Network. Available at: <https://londonminingnetwork.org/wp-content/uploads/2017/09/Fundao-Report-Final-lowres.pdf>.

Rodrigues, L. (2019a). 'Decisão impede que réus na tragédia de Mariana respondam por homicídio'. [online] Agência Brasil. Available at: <https://agenciabrasil.ebc.com.br/justica/noticia/2019-04/decisao-impede-que-reus-na-tragedia-de-mariana-respondam-por-homicidio>.

Rodrigues, L. (2019b). 'Acordo prorroga auxílios pagos a atingidos por tragédia de Brumadinho'. [online] Agência Brasil. Available at: <https://agenciabrasil.ebc.com.br/justica/noticia/2019-11/acordo-prorroga-auxilios-pagos-atingidos-por-tragedia-de-brumadinho>.

Rodrigues, L. (2021a). 'STJ suspende processo que pedia extinção da Fundação Renova'. [online] Agência Brasil. Available at: <https://agenciabrasil.ebc.com.br/justica/noticia/2021-05/stj-suspende-processo-que-pedia-extincao-da-fundacao-renova>.

Rodrigues, L. (2021b). 'Auxílios pagos pela Vale irão virar programa de transferência de renda'. [online] Agência Brasil. Available at: <https://agenciabrasil.ebc.com.br/justica/noticia/2021-02/auxilios-pagos-pela-vale-irao-virar-programa-de-transferencia-de-renda>.

Rodrigues, L. (2021c). 'Revelamos o que diz o laudo sigiloso da PF sobre a tragédia em Brumadinho'. [online] Agência Pública. Available at:

<https://apublica.org/2021/09/revelamos-o-que-diz-o-laudo-sigiloso-da-pf-sobre-a-tragedia-em-brumadinho/>.

Rodrigues, L. (2021d). 'Brumadinho dam collapse now two years old'. [online] Agência Brasil. Available at: <https://agenciabrasil.ebc.com.br/en/geral/noticia/2021-01/brumadinho-dam-collapse-now-two-years-old>.

Século Diário (2019) 'MPF constata: apenas um dos 42 programas da Fundação Renova tem sido cumprido'. [online] Available at: <https://www.seculodiario.com.br/meio-ambiente/mpf-constata-apenas-um-dos-42-programas-da-fundacao-renova-tem-sido-cumprido>.

Servín, R. (2014). 'Analizan más demandas contra GMéxico; es el peor desastre ambiental, afirma la Semarnat'. [online] *El Financiero*. Available at: <https://www.elfinanciero.com.mx/sociedad/analizan-mas-demandas-contra-gmexico-es-el-peor-desastre-ambiental-afirma-la-semarnat/>.

Tokarnia, M. (2020) 'Tragédia de Mariana faz 5 anos e população ainda aguarda reparações'. [online] yahoo! notícias. Available at: <https://br.noticias.yahoo.com/tragédia-mariana-faz-5-anos-235300249.html>.

Tribunal de Justiça do Estado de Minas Gerais (TJMG) (2021) 'Presidente do TJMG homologa acordo histórico entre a Vale e instituições públicas'. [online] Available at: <https://www.tjmg.jus.br/portal-tjmg/noticias/presidente-do-tjmg-anuncia-acordo-historico-entre-a-vale-e-as-instituicoes-publicas-8A80BCE676728EAA01776D32461E7848.htm#.Yg493S-l1pQ>.

UOL (2021) 'Fundação Renova não entrega reassentamento de casas destruídas em rompimento de barragem em Mariana'. [online] Available at: <https://cultura.uol.com.br/noticias/17039_fundacao-renova-nao-entrega-casas-destruidas-em-rompimento-da-barragem-em-mariana.html>.

Vega, A. (2019) 'Así es la vida en las comunidades del Río Sonora, a 5 años del derrame tóxico de Grupo México'. [online] Animal Político. Available at: <https://www.animalpolitico.com/2019/08/comunidades-derrame-toxico-grupo-mexico/>.

Viana, V. and Cobuci, J. (2021) 'A luta dos atingidos da bacia do rio Doce que tiveram o Auxílio Financeiro Emergencial cancelado/diminuído pela Fundação Renova'. [online] *Jornal A Sirene*. Available at: <https://jornalasirene.com.br/justica/2021/03/02/a-luta-dos-atingidos-da-bacia-do-rio-doce-que-tiveram-o-auxilio-financeiro-emergencial-cancelado-diminui-do-pela-fundacao-renova>.

Weimann, G. (2020) 'Brumadinho: Vale tenta boicotar assessorias técnicas dos atingidos pelo rompimento'. [online] *Brasil de Fato*. Available at: <https://www.brasildefato.com.br/2020/01/20/brumadinho-vale-tenta-boicotar-assessorias-tecnicas-dos-atingidos-pelo-rompimento>.

CHAPTER 6
Water: The industry's Achilles heel

Mining has severe impacts on both the quantity and quality of water. Large mines can consume millions of litres of water a day, putting immense pressure on water supply for communities. Mineral processing also contaminates vast amounts of water, which may become a health and environmental hazard unless it is properly treated and contained. This chapter looks at these two aspects of the issue, focusing on illegal mining (garimpo) in the Brazilian Amazon; and industrial copper mining in Chile, in the context of the country's ongoing megadrought.

Mining is a thirsty industry. Mines require uninterrupted water access for mineral processing, tailings ponds, dust suppression, and slurry transport, not to mention the needs of their employees (Kuyek, 2019, p. 45). Large mines can consume hundreds of litres of water a second, amounting to tens of millions of litres every day.

This thirst is likely to become ever more insatiable in the coming years. As the mining industry steadily exhausts the best mineral deposits, ore grades – the concentration of the desired material in the rock that contains it – are declining. This means that processing the ore requires ever greater volumes of water to produce a tonne of metal (Prosser et al., 2011, p. 137).

Without access to vast quantities of water, industrial mining would almost be impossible. But as many mines are located in areas where water is already scarce, conflict with local communities is inevitable.

In Latin America, these conflicts are being exacerbated by climate change, which is already having major impacts on water supply. Over the last two decades, droughts have ravaged Mexico, Central America, Bolivia, Brazil, Chile, and even the Amazon, the world's largest tropical rainforest. Some of these droughts are the most severe on record.

The problem is not simply that the mining industry consumes a great deal of water, but that it often contaminates existing water sources, rendering them useless for nearby populations unless comprehensive and long-term water treatment programmes are put in place.

Mines create sediment which gets carried into rivers and lakes, polluting the water and smothering aquatic ecosystems. There is a risk of acid mine drainage, a serious long-term environmental problem which may continue to poison downstream waters long after any mining activity has ceased. And miners use toxic chemicals such as cyanide, sulphuric acid, and mercury to process ores, which may get spilt, or leach from mine sites into lakes and rivers (Kuyek, 2019, pp. 39–43).

Virtually wherever there is conflict between a community and a mine, one of the principal grievances will be water. This chapter will examine these two aspects of the problem – pollution and scarcity – looking at the problem of wildcat mining or *garimpo* in the Brazilian Amazon and the impact of industrial mining on water supply in Chile.

I. Assimilate or die: *garimpo* in the Brazilian Amazon

'Where would Brazil be today, had it not been for the *bandeirantes* who exploited the diamonds?' asked former Brazilian president Jair Bolsonaro in 2017, while still a deputy in Congress. 'We would only have a third of our current territory if it hadn't been for them.' (Audi, 2018)

The *bandeirantes* were explorers, slavers, and fortune hunters, who led expeditions into the Brazilian interior from São Paulo from the 16th to the 18th century. Their aim was to capture and enslave native populations and search for gold, silver, and gemstones. In the process, they founded settlements and expanded the borders of the Portuguese colony well beyond the limits that had been agreed with Spain in the Treaty of Tordesillas.[1]

Later, during the 19th and early 20th centuries, Brazilian artists, statesmen and intellectuals – particularly from São Paulo – constructed a powerful mythology around the *bandeirantes*, championing them as heroes and founders of the nation.

Since the 1960s, with growing awareness of issues of race and racism in Brazil, historical interpretations of the *bandeirantes* have become more critical, a shift due in large part to activism by Afro-Brazilian and indigenous movements. More recently, the international Black Lives Matter movement has fed into this debate, with an ongoing dispute over the many statues and monuments to the *bandeirantes* analogous to that rumbling on in the UK over statues of figures such as Edward Colston, Cecil Rhodes, and others.

Nonetheless, the mythology around the *bandeirantes* remains influential and so, to this day, the artisanal miner or *garimpeiro* is celebrated as a romantic figure, someone able to survive and prosper in an alien and hostile environment through his own resilience, resourcefulness, and hard work. In this sense, *garimpo* is an activity deeply engrained within traditional Brazilian – and especially *paulista* – notions of identity.

Early miners in Brazil largely worked alluvial deposits, first in São Paulo, then in Minas Gerais in the famous 'gold cycle' of the 17th century (Lopes, 2018). The first *garimpeiros* were miners who worked without authorization from the Portuguese Crown (Brown, 2012, p. 84). But while the traditional image of a *garimpeiro* is one of a lone man panning for gold, knee-deep in a river, in recent decades this has been superseded. What comes to mind instead are areas of once pristine rainforest, scarred by craters full of stagnant water.

Garimpo today is big business, a semi-industrial activity involving the use of expensive heavy machinery: excavators, dredgers, tractors, light aircraft,

Photo 6.1 *Garimpo* in the Munduruku Indigenous Territory, Pará / Vinícius Mendonça/IBAMA 2018 / CC BY-SA 2.0

even helicopters. While the Brazilian Constitution does permit *garimpo*, the controls to which the industry is subject, along with the presence of lucrative mineral deposits in areas where mining is forbidden – such as indigenous territories and national parks – mean that most *garimpo* sites in Brazil today operate outside the law (Minas Júnior Consultoria Mineral, 2019).

With the current system for registering gold from small-scale mines being unfit for purpose, it is easy for *garimpeiros* to launder their gold by registering it under a false license, so it appears to have come from a site where mining was permitted. Unlike cocaine, which is illegal in any form, once laundered, illegal gold is indistinguishable from that mined legally. After it has disappeared into global supply chains, it is virtually impossible to trace (Jordan, 2021). Annual turnover from illegal mining in Brazil varies between R$3 billion [$570 million] and R$4 billion [$760 million] (Nikou and Moncau, 2021a).[2]

In short, *garimpo* today is high-level organized crime, with links to other criminal enterprises such as drug and gun running; money laundering; and human trafficking, mostly of women and girls coerced into sex work.

Minamata revisited

Jorge Bodanzky is a filmmaker who is currently working on a film about the effects of *garimpo* in the Amazon entitled *The Amazon: a new Minamata?* One of his main subjects in the film is Erik Jennings, a neurosurgeon at the public hospital in Santarém, in the Amazonian state of Pará. Jennings works for SESAI, Brazil's dedicated indigenous health agency, and has worked extensively amongst indigenous communities in the Tapajós river basin.

'Dr Erik and his colleagues were worried by the significant increase in the number of requests for wheelchairs for schoolchildren in the area. These were children who were suffering from neurological problems and issues with their movement and coordination,' says Bodanzky. 'There were also signs of a very high rate of failure in the schools, students being held back and having issues with their studies, way above the average rate for the region.'

Jennings' hypothesis? Mercury poisoning.

Mercury – one of ten chemicals identified as a major public health concern by the World Health Organization (WHO, 2020) – has been used for centuries as an inexpensive way of refining gold. It is indestructible and can be carried for hundreds of miles by water and wind, meaning that once released into the environment it is almost impossible to contain (Crespo-Lopez et al., 2021).

In *garimpo* mercury works like a magnet, binding with the gold particles and making them easier to separate from the surrounding ore. This amalgam is then washed to separate any remaining waste, which is dumped in the river. The Brazilian Federal Police estimate that *garimpeiros* throw around seven million tonnes of sediment into the Rio Tapajós per year, much of which contains mercury (Nikou and Moncau, 2021a).

To extract the gold, the *garimpeiros* typically burn off the mercury from the amalgam. The vapours released partially condense in the clouds and are then washed into water sources by the rain, where the action of methylating bacteria turns the mercury into methylmercury, a powerful neurotoxin which bio-accumulates in fish and bio-magnifies up the food chain. On top of this, the soil in the Amazon is naturally high in inorganic mercury. Though this presents no risk to human health when the soil is undisturbed, *garimpo* – along with other human interventions which remove forest, such as cattle ranching – causes erosion, sending this soil into the water, where again, the mercury becomes methylmercury.

Methylmercury attacks the central nervous system, causing symptoms such as memory loss, lethargy, and problems with concentration – symptoms which can be mild at first and are often misdiagnosed. Eventually – in some cases years later – more serious complications may develop: tremors, poor hearing and speech, severe cognitive impairment; in the most extreme cases, paralysis, coma, and death. The damage is irreversible (Console, 2019).

It can also cross the placenta, having severe toxic effects on the brain and nervous system during embryonic development. Unborn babies are particularly susceptible to methylmercury and may be affected even if the mother displays no signs of poisoning (Minai, 2018).

Bodanzky's film draws a parallel with the Minamata Disaster in Japan, when thousands of people were poisoned after consuming fish and shellfish contaminated with mercury from the Minamata Bay. A factory owned by the Chisso Corporation was identified as the culprit – it dumped wastewater in the bay for decades, continuing to do so even after the link between its pollution and the so-called 'Minamata disease' had been established (Emmet Hernan, 2022).

'Everything that happened in Minamata has started happening in the Amazon,' says Bodanzky. 'On a different scale, but we can see what might happen if nothing is done about it.'

'They're killing us'

The problem is particularly acute in the communities which live along the rivers, including the Munduruku, one of Brazil's largest indigenous groups. In recent years, they have come under increasing pressure from a range of different sectors, including illegal loggers, land grabbers, agribusiness, and hydroelectricity generation.

However, *garimpo* – which has been a threat to the Munduruku intermittently since the 1970s – exploded during Bolsonaro's presidency. Research by the Instituto Socioambiental found that the area of the Munduruku Indigenous Territory degraded by *garimpo* increased 363 per cent between January 2019 and May 2021 (Aragão, 2021). The impact on the Rio Tapajós – on which the Munduruku depend for fishing, feeding livestock, bathing, and transportation – has been especially devastating.

'We know that we're sick, that we've got levels of mercury in our bodies. But this is the only source of food we have,' says leader Alessandra Korap Munduruku. 'We can't stop eating fish because it's contaminated with mercury. If we did, we would all starve because we depend on the river, we depend on the fish for our survival.' (Console, 2019)

The level of mercury intoxication amongst the Munduruku surpassed even the worst expectations of Erik Jennings and his team, who have been monitoring the situation closely in recent years.

'Our research [involving 109 residents of the Alto Tapajós region] shows that 99 per cent of the population tested have levels of mercury in their blood which surpass the threshold considered safe by the World Health Organization,' Jennings told the investigative journalism platform Repórter Brasil. 'Some of them are 15 times over the limit. It's extremely worrying.' (Nikou and Moncau, 2021b)

'There has to be some punishment, because they're killing us,' says Alessandra Munduruku. 'There can be no question about whether *garimpo* will have an impact; it already has done. The fish are dying from the contamination. The women are being poisoned. Now we've got proof.' (ibid.)

In late 2020, Jennings attempted to return to the villages of the Alto Tapajós with Bodanzky's producer and cameraman to conduct the handover of these results with the Munduruku. However, when their plane stopped to refuel in Jacareacanga, one of the towns bordering the indigenous territory, they were attacked by a group of pro-*garimpo* indigenous people.

'They thought that our guys were from Greenpeace and a big argument broke out, it got quite ugly,' explains Bodanzky. 'One of the indigenous

Photo 6.2 Alessandra Korap Munduruku / Amazônia Real 2021 / CC BY-NC-SA 2.0

women started to paint them red – a sign of war. Our producer Juliana got all covered in red and they were really scared about what might happen.'

Jennings and the team escaped, managing to restart the plane and take off amid a hail of stones. Miraculously, no-one was hurt.

A garimpeiro *president*

The episode is a graphic illustration of the increasingly bitter divisions that have emerged in indigenous communities in recent years.

'There are indigenous people and even some indigenous leaders who say that mercury isn't an issue at all, that's all a nonsense, and that what's important is getting access to the gold,' says Bodanzky. 'The miners make them promises – cars, money, mobile phones, women – this co-opts them and creates division within indigenous communities.'

Such divisions can be seen not only in indigenous communities affected by *garimpo*, but also in those living close to agribusiness interests. There are indigenous communities elsewhere in the Amazon which use lands on their reserves to farm monocultures such as soy and corn using modern methods – including GM crops – and partnering with outsiders.

Though this is illegal, it has been encouraged by local politicians and particularly by Bolsonaro's government, which championed these pro-mining and pro-agribusiness indigenous communities as models for Brazil's indigenous population in the 21st century. It's a cynical game of divide and rule, with the aim of dismantling the legal protections afforded to indigenous communities and giving outside interests access to the abundant natural resources contained within their territories.

For example, in August 2020, Ricardo Salles, then minister of the environment, had a Brazilian Air Force plane fly a group of seven pro-*garimpo* indigenous people from Jacareacanga to Brasília for a meeting. The plane had been intended for use in operations to combat environmental crime, particularly *garimpo* (Carneiro, 2020). Under any normal administration, Salles' decision to illegally commandeer a military plane to transport a group of suspected environmental criminals might have seen him removed, or worse. But Bolsonaro's was no normal administration, and until his resignation in June 2021 – after becoming a suspect in two Federal Police investigations into illegal logging in the Amazon – Salles had been one of his closest allies.

Bolsonaro had been a staunch defender of *garimpo* since long before he became president. His father, Percy Geraldo Bolsonaro – an unlicensed dentist in rural São Paulo – was one of the estimated 100,000 *garimpeiros* who participated in the famous Serra Pelada gold rush of the 1980s, in the east of Pará. The subject of a much-celebrated series of photos by Sebastião Salgado, the Serra Pelada *garimpo* was the largest open-air goldmine in the world, to which men flocked from all over Brazil to dig for gold in atrocious conditions.

Many were farmhands who had been booted off their land in the neighbouring state of Maranhão. While a lucky few became rich, most returned to their hometowns just as poor as when they left. Some never returned at all, the victims of accidents, murder, or suicide; or succumbing to disease – there were outbreaks of malaria, leprosy, and AIDS in the Serra Pelada camps. To this day, the number of men who died at the Serra Pelada *garimpo* is unknown (Kotscho, 2021).

Serra Pelada was viable for manual mining only until late 1989, and now stands abandoned. Nonetheless, there are still *garimpeiros* in the area who believe there is gold beneath the old pit and in 2018 they delivered a petition to Bolsonaro asking for the end of regulations which prevent them from digging for it (Audi, 2018).

For today's *garimpeiros* then, Bolsonaro is one of their own. 'The whole Transamazônica region and the areas where there is *garimpo* are Bolsonaro through and through,' says Bodanzky.

'Colonization has never stopped'

For Jennings, the poisoning of the Munduruku and other riverside communities in the Amazon is not simply a public health issue. In Bodanzky's film he calls it 'the tip of the iceberg'.

'The question of public health is just one dimension of a much bigger problem. At its core, the issue is a political one and it's about the way in which the Amazon is being occupied and used,' Bodanzky says. 'Essentially it's a political issue, and the political situation in Brazil at the moment is catastrophic.'

Under Bolsonaro, deforestation in the Brazilian Amazon surged to its highest level in more than a decade. The organs of the Brazilian state responsible for the protection of the environment and the country's indigenous population – such as the environmental enforcement agency Ibama and the National Indian Foundation (Funai) – were crippled, with budgets slashed and personnel laid off.

'From day one, Bolsonaro has signaled he would prioritize economic growth over any form of restraint or care for the environment and forest peoples,' says Christian Poirier, programme director at the NGO Amazon Watch (Volckhausen, 2020).

In essence, Bolsonaro is in thrall to the old dream of colonizing the Amazon. He sees the forest as territory to be conquered, its abundant natural resources just value to be extracted. Its inhabitants, particularly the indigenous, must assimilate or die.

This vision is clear not only in his tributes to the *bandeirantes*, but in his admiration for Brazil's military dictatorship (1964-85), which, with its Operation Amazon, built 10,000 miles of roads in the region in seven years and provided generous incentives and tax breaks to extractive industries and agriculture. Like Bolsonaro, its policy towards the indigenous was one of integrationism – in other words, they were not to be left to pursue their traditional lifestyles free from outside interference, but were to be integrated into mainstream Brazilian society as quickly as possible (Comissão da Verdade do Estado de São Paulo, 2015).

Many of those who resisted were killed: according to the report produced by the National Truth Commission – set up in 2011 to investigate state human rights abuses between 1946 and 1988 – at least 8,350 indigenous people were killed during this period, either directly by state agents or as a result of state neglect. State violence towards the indigenous was particularly intense from December 1968, following the AI-5 decree which closed Congress and tightened the military's grip on power, signalling the beginning of the so-called 'Years of lead' – the most repressive phase of the dictatorship (Dias et al., 2014, pp. 204–205).

'Colonization has never stopped,' says Alessandra Munduruku. 'It's only advancing further' (Brazil Matters, 2021).

Bolsonaro narrowly lost the presidential election in 2022 to his political nemesis Luiz Inácio Lula da Silva, and at the time of this book going to print, had recently fled Brazil for Florida, reportedly fearful of prosecution. Lula now faces an enormous challenge to repair the damage done to the Brazilian environment in the last few years, including the closure of illicit *garimpos* dotted all over the Amazon. This won't be easy, and nor will it be cheap. Thousands of miners will have to be removed and integrated into legitimate sectors of the labour market.

Some have suggested that the solution lies in allowing industrial mining companies to move in. After all, even if they sometimes flout the rules, mining companies have to submit Environmental Impact Assessments, abide by local environmental regulations, and provide some return to the state in the form of taxes and royalties. But industrial mining is no more compatible with the traditional subsistence-based cultures of the native people of the Amazon than *garimpo*. It has already caused conflicts with indigenous communities in the region, such as the Xikrin, who for years have denounced the contamination of a river running through their territory by the Onça Puma nickel project, belonging to Brazilian mining giant Vale (APIB and Amazon Watch, 2021, p. 39).

There is also a link between *garimpo* and industrial mining: as they move further into indigenous lands, the *garimpeiros* are breaking down the social, political, and legal barriers to the operation of larger mining enterprises. It is no surprise that the recent advance of *garimpo* has coincided with a spike in the number of applications to mine on indigenous lands by legitimate mining companies, with the highest number registered in 2020 for 24 years (Jordan and Potter, 2021).

One such company is the British multinational Anglo American, which, as recently as late 2020, had 13 prospecting and research applications overlapping with Sawré Muybu, another traditional Munduruku territory further to the north (Miranda, 2021a). Although mining research on indigenous land remains illegal, Sawré Muybu may be more vulnerable, as unlike the Munduruku Indigenous Territory, the Brazilian state has yet to ratify its demarcation.

Following months of pressure from indigenous and environmental groups, Anglo American withdrew 27 applications overlapping with indigenous lands in Brazil in July 2021, including those affecting Sawré Muybu (Miranda, 2021b). Nonetheless, the company has refused to rule out any future mining activity on indigenous land (Amazon Watch, 2021).

When it comes to water, this does not bode well, as communities in Chile can attest.

II. Mining in a megadrought: Anglo American in Chile

'The older people waited close to the well, while we younger ones created a distraction by blockading the road that leads to the mine. The idea was to get all the police to come and deal with the trouble that we were making,' says Ximena Gallardo, an environmental engineer and activist with the group Poyewn, from El Melón, in the Valparaíso region of central Chile.

'It worked: when the police arrived, they all made straight for the roadblock. So the older people broke into the well and occupied it.'

The well in question belonged to Anglo American, which has been the majority owner and operator of the nearby El Soldado copper mine since 2002. El Soldado has 16 deep water wells and 19 permits, giving it access to 453 litres of water per second, though environmental regulations mean it is currently limited to 120 l/s (Santibáñez et al., 2020). Anglo claims it has reduced its consumption further still, to 109 l/s, but given that the mine operates continuously this still works out at more than 9.4 million litres of water a day (Marconi and Urquiza, 2020).

In contrast, El Melón – a town of 9,000 people – has had access to just eight litres a second since 2019, though at least 30 l/s are required to ensure constant domestic supply. Some residents complain they have had no running water at home since 2015.

In October 2019, residents sent a petition to Anglo American with a list of 24 demands. The company ignored them (ibid.). So on 7 November, Gallardo and the other locals occupied the well.

'When Anglo American showed up all of us who were up by the mine came down to support the older ones,' she says. 'We set up a camp and moved in. We put a padlock on the valve and we said we wouldn't give them any more water until part of it had been redirected to the municipality.'

The occupation lasted for three months until the police removed the camp in February 2020. It succeeded in getting Anglo American to agree to redirect 20 l/s of the water from the well to the municipal network. However, with the municipal government – perceived to be too close to the company – responsible for overseeing this, Gallardo is sceptical as to whether Anglo American is being held to its promise.

In June 2020, at the height of the first wave of the Covid-19 pandemic in Chile, Gallardo and 12 others filed an injunction against Anglo American, requesting that El Melón be guaranteed enough water to comply with basic hygiene precautions. While this request was initially rejected by a court in Valparaíso, the Supreme Court reversed this decision in January 2021, ruling that 100 litres a day should be guaranteed to every member of the community.

However, it was only a partial victory: though the complaint had been made against Anglo American, responsibility for providing water to the community was assigned to the local government. In fact, the Supreme Court stressed that Anglo American bore no responsibility for the water shortages and cleared the company of any wrongdoing.

And the conflict is far from being resolved. The main problem is that just 30 per cent of the local population are connected to the municipal water network. The rest, particularly those in rural areas outside the town or on higher ground, depend on wells which continue to provide only a very limited supply.

'Anglo American affects 100 per cent of the population in one way or another, but the changes they've made only benefit 30 per cent. And the worst of it is that they say they're the result of voluntary agreements, when in fact they're the result of our roadblocks, occupations, and protests,' says Gallardo, with obvious anger. 'It's almost a double-edged sword for us, because everything that we achieve through putting pressure on them, they then use to whitewash their image.'

A land without water

'Not having water means you can't satisfy your basic needs. You can't drink water. You can't wash your dishes. You can't use the bathroom,' says Gallardo. 'It's a luxury to have a hot shower. We have to wash with cold water from a washpot, because our showers have no pressure.'

El Melón is not the only town in Chile undergoing such hardships. According to a survey by the Chilean Ministry of Social Development from 2017 (the most recent year for which figures are available), more than 1.4 million people in Chile did not have access to drinking water at home (Gobierno de Chile, 2019), with that figure likely to be higher still today.

Summer droughts are common in central Chile, which enjoys a sunny, Mediterranean climate of hot summers and mild winters. 'In El Melón it's a given that in summer there won't be any water,' says Gallardo. 'People expect the wells to dry up in the summer, so in the winter they stock up.'

But these seasonal droughts have been increasingly affecting the winter months as well. Since 2010, rainfall in central Chile has been below normal by an average of 20 to 45 per cent (Voiland, 2020), and in 2019, the year the

Photo 6.3 The daughter of mining opponents in the dry landscape around the El Soldado mine site / © Elena Rusca 2020

residents of El Melón occupied Anglo American's well, the rainfall deficit was as much as 80 to 90 per cent (Voiland, 2019).

The situation is frequently described as a 'megadrought', and it is the country's worst on record. Scientists estimate that the last time there was a drought on such a scale was over 1,000 years ago. While climate change is a factor, most of the intensity and severity of the drought is attributed to natural cycles in atmospheric circulation, rainfall patterns and ocean temperatures (Voiland, 2020). To make matters worse, the drought has been accompanied by exceptional heatwaves and wildfires in recent years.

Decrepit infrastructure has also played a part in the shortages, with almost 35 per cent of urban drinking water thought to be lost to leaky pipes (Guzmán, 2019). But while Anglo American and other large water consumers blame the drought and poor infrastructure for the shortages, this wilfully overlooks Chile's system of water management, which, since the early 1980s, has prioritized water for agriculture and industry over domestic use.

'For sure, these factors [climate change and poor infrastructure] have exacerbated the situation, but none of these industries can hide behind climate change when the little water that is available is being directed preferentially towards their productive activities rather than for human consumption,' says Matías Asun, the director of Greenpeace Chile (Greenpeace Chile, 2020).

The Chilean model

Chile is a unique case in that it has gone further than virtually any other country on the planet in the privatization and commodification of water.

The private ownership of water was established in the Constitution of 1980, written and approved at the height of General Pinochet's brutal 17-year dictatorship and consolidated in agreements that allowed a transition to a civilian-elected government in 1990. A Water Code in 1981 built on this, separating water rights from land ownership – meaning that a landowner does not necessarily have the right to draw water from a river flowing through their property. In contrast, the Mining Code of 1983 did grant mining companies the rights to any water found within a given concession.

The state was given responsibility for granting water rights, which are held in perpetuity and are traded in a free marketplace with little state oversight. Today, the Chilean state does not know who holds water rights, nor how they are using them. The results of this radical experiment are extreme: today in Chile, 1 per cent of the 29,001 entities which have rights to extract and consume water control more than 79 per cent of the total water available in the country (El Desconcierto, 2020). For example, the Walker Prieto family – a prominent establishment clan with a long history at the highest levels of business and politics – has rights amounting to 29,000 litres of water a second. It is estimated this would supply 17 million people for domestic use (CIVICUS, 2020).

Water rights also appear to have been handed out with little thought to the quantity of water the Chilean environment can provide. The quantity of water covered by existing rights currently exceeds the amount of available water threefold, while in the Valparaíso region, it's sevenfold (*El Observador*, 2020). This has led to some outrageous examples of speculation. In 2019, to obtain more water for rural communities in the province of Petorca – not far from El Melón – the Ministry of Public Works bought the rights to eight litres of water a second from an agribusiness enterprise for 184 million pesos (around $239,000), though this same company had bought the rights a year beforehand for just 40 million – and this is far from the only such example. In other words, faced with an acute water crisis, the Chilean state is paying a premium for water which was once under its control and which it gave away for free to private interests (Arellano, 2020).

Many communities now depend on water deliveries, again, at considerable cost to the Chilean taxpayer: from 2011 to the end of 2019, the Chilean state paid at least 222 billion pesos [roughly $290 million] to private service providers for deliveries of water, mostly via tanker lorries, in 15 out of Chile's 16 regions. With the water crisis as a justification, many of these contracts have been agreed directly between the state and service provider, without proper bidding processes. The values negotiated are often arbitrary and opaque, with huge variations depending on when and where the service was contracted (Ameonna, 2020).

'We're delivering water to communities via tankers for their domestic supply. But there is water in these river basins; it's just that human consumption isn't given priority,' says Lucio Cuenca, an engineer and director of the Latin American Observatory of Environmental Conflicts (OLCA). 'What takes priority instead is mining, agribusiness, avocados, lemons, vineyards. This shows you how confused we've got our priorities in Chile.' (MAT, 2019)

Indeed, El Melón and other communities in the area have to compete for the scarce water resources not only with El Soldado, but with agribusiness, particularly large plantations of avocados and citrus fruit, much of which is destined for export to global markets. One local agricultural enterprise has rights to nearly 695 l/s, well above Anglo American's 453 l/s (Santibáñez et al., 2020).

I asked Ximena Gallardo why the residents of El Melón have singled out El Soldado when there are other interests in the region that may be consuming even greater quantities of water.

'You might be paying for rights without necessarily using them,' she says. 'There needs to be a clearer distinction between the rights that someone is paying for and the actual amount of water they consume.'

And not all water consumption has the same outcome. 'Water use in mining is intensive: the water is used 24 hours a day, during which time it is contaminated by the mining processes and is good for nothing afterwards,' says Cuenca. 'In agriculture, while part of the water is spent on irrigation, another part returns to the natural cycle.' (MAT, 2019)

Chile's water crisis has taken on further urgency since the outbreak of the Covid-19 pandemic, and in August 2020, the UN's special rapporteur on the human rights to drinking water and sanitation criticized Chile for failing to prioritize the rights to water and health of its population over those of business.

'We've had Covid in El Melón,' says Gallardo. 'And we were bombarded with information about how we had to wash our hands for 20 seconds. But we had no water.'

'We almost got buried alive'

The anger of Gallardo and the other residents of El Melón towards Anglo American goes beyond the water shortages. There is also much concern about the El Torito tailings dam, located just 7.5 kilometres from the town.

Chile is one of the most seismically active countries on the planet. Two of the six biggest earthquakes on record took place in Chile, and minor tremors are constant. In 1965, an earthquake at the town of La Ligua, 26 kilometres north of El Melón, caused two tailings dams belonging to El Soldado to fail. The subsequent avalanche of mining waste wiped the nearby village of El Cobre off the map, killing at least 200 people, though the true total is likely to be much higher. Most of the bodies were never recovered.

El Torito's current capacity is around 75 times the total volume of tailings released in the 1965 disaster (Dobry, 1965, p. 86). According to a study by the consultants Golder, were El Torito to fail, most of El Melón would be buried beneath mine tailings in a matter of hours (Marconi and Urquiza, 2020).

'Whenever there's a tremor – because we get a lot of them here – the first thing people do is look up at the dam to see if it's going to break,' says Gallardo. 'People are scared. There are children of 12 who've inherited this fear from their grandparents.'

This fear is fuelled not only by memories of the 1965 disaster, but by more recent events. In 2007, following a tremor, Anglo American said that a crack had appeared in the containing wall of the tailings dam. Despite this, in 2019 the company was granted permission to expand El Torito by another 1.4 metres a year until 2027, increasing the dam's capacity to 235 million tonnes – an increase of 30 per cent (Fundación Terram, 2018).

'We almost got buried alive and now the authorities give the company permission to build it even higher,' says Jorge Ramírez, a local electrical technician. 'It's like having a 100-storey building which is collapsing and you want to build it another 30 storeys higher. It's completely illogical.' (Marconi and Urquiza, 2020)

On top of this, the dam has long had a seepage problem, with tailings water seeping into the aquifer. To contain this, Anglo American installed a 'hydraulic barrier': a filtration and drainage system designed to prevent contaminants reaching the community's drinking water.

'They've installed drains, holes in the ground where they pump out the contaminants and return them to the dam,' explains Gallardo. 'But this is all based on statistical modelling, projections. It's not real. There aren't any follow-up studies which say that these measures are working.'

The evidence suggests her scepticism is justified. In 2014, Anglo American was hit by a $4.5 million fine by Chilean environmental authorities for violating 16 regulations, with one of the problems highlighted being the high levels of sulphates detected in several wells in the region (Portal Minero, 2014). Subsequently, a report published in 2019 by Fundación Relaves, a Chilean NGO which monitors tailings storage facilities in the country, found levels of manganese, iron, and sulphates in the groundwater in El Melón beyond permitted levels, and concluded that sulphate levels are increasing due to the gradual filtration of tailings water into the aquifer (LMN, 2020).

'They only put these kind of mitigation measures in place because the law says they have to. It doesn't matter whether they actually work or not,' says Gallardo. 'These measures don't help the community. They're not focused on the community. The focus is on avoiding fines or making sure that they're as low as possible.'

Glaciers in danger

Anglo American's activity doesn't only threaten rural communities such as El Melón; its plans to expand Los Bronces, another major copper mine, have even alarmed residents in Santiago, Chile's capital and largest city.

Located in the *cordillera* only around 35 km from Santiago's outer suburbs, Los Bronces has been worked since the 19th century, and the ore quality is declining. To ensure ongoing viability, Anglo's proposed solution is to dig what is effectively a new, underground mine, five kilometres from the project, giving it access to ores of a mineral concentration three times those found in the current pit (Jamasmie, 2019). Anglo also wants to expand the existing pit on two sides, as well as increase the water recirculation of the project. The required investment is estimated at more than $3 billion.

But Los Bronces is located in an extremely delicate mountain ecosystem. The planned underground section overlaps with the Yerba Loca nature reserve and the project is located within just a few kilometres of several glaciers. More than 70 per cent of Chile's population depend on glacial runoff for their water supply and given Chile's current troubles with drought, protecting the glaciers is becoming an increasingly urgent political issue. Critics argue that the proposed expansion could endanger the water supply for Santiago and the surrounding metropolitan area – a region home to over seven million people, more than 40 per cent of the Chilean population.

Anglo American claims that these fears are unfounded and all the publicity for the expansion states explicitly that there will be no impact on glaciers. But since the early 1990s, various studies have documented the damage mining has done to Chile's glaciers – including by Los Bronces.

Mining affects glacial ecosystems in various ways. It generates dust and pollution which settle on glaciers and accelerate ice melt. Vibrations from blasting can damage glacial areas. Companies use glaciers as deposits for ballast and debris. And sometimes they simply remove glacial ice outright where it presents an obstacle to extraction, as in the case of the Chilean state copper miner CODELCO, which removed glaciers in the 1970s to make way for its División Andina mine (Millan Lombrana et al., 2019).

In the case of Los Bronces, glaciologists Alexander Brenning and Guillermo Azócar found that of the 1.9 km² of rock glaciers in the area, mining activity had destroyed 0.8 km² by as early as 1997, resulting in the permanent loss of between six and nine million cubic litres of fresh water (Brenning and Azócar, 2010).

While Anglo stresses that mining beneath Yerba Loca means that there will be no impacts on the surface, it will have to build chimneys to ventilate emissions by diesel-powered vehicles working underground day and night to remove raw material from beneath the reserve. Given the proximity of the glaciers, it is hard to see how it can make such a guarantee in good faith.

'The tunnel will have chimneys along it in certain places, which will emit particulate material,' says Felipe Espinosa, the director of the Chile Glaciers Foundation. 'This material will land on glaciers, damaging them. Besides which, the tunnel will run beneath the nature reserve and we've no way of knowing whether that will have an impact.' (Surf Beats Radio, 2020)

The future of the project now looks uncertain. In May 2022, Chilean environmental regulators rejected Anglo American's application for the expansion of

Los Bronces, claiming that the company had not provided enough information on potential risks to public health. Anglo filed an appeal the following month and says it remains committed to the project.

Building a 'glacial country'

Unlike neighbouring Argentina, Chile still does not have a specific law protecting its glaciers, though it is home to around 80 per cent of the glaciers in South America. There have been repeated attempts to pass this legislation going as far back as 2005, but none has been successful, with the mining sector using its considerable influence either to block or to water down any proposals it feels are detrimental to its interests. This led Greenpeace and other environmental organizations to *oppose* a glaciers law drawn up during Michelle Bachelet's second term (2014-18), arguing that rather than protect the glaciers, the law established the legal conditions for their destruction.

Victoria Uranga is a journalist and member of various environmental organizations, including No+Anglo, a network of communities in Chile impacted by Anglo American.

'The mining lobby is very strong and they always say that if the glaciers are given more protection, they will have to reduce their operations and that will have huge impacts on the Chilean economy,' she says. 'In the context of the pandemic that's very threatening. People are vulnerable, they're in situations of poverty, there's unemployment. It's as if that were all directly linked with mining.'

The glaciers are important not only as a source of fresh water for the people of Chile. Their white surfaces reflect heat back into space, helping to regulate the global climate. They also provide scientists with a record of how the climate has changed over time; today's rapidly melting glaciers are one of the clearest indications of anthropogenic climate change.

Like the melting ice caps at the poles, glacial melt is contributing to rising sea levels. The deposit of large quantities of fresh water in the sea disrupts ocean currents, which in turn alters marine ecosystems and changes weather patterns. Alpine glaciers such as those in the Andes are particularly vulnerable: they are melting faster than the ice sheets in Greenland and the Antarctic and are thought to be the largest contributor to rising sea levels in the 21st century, after the thermal expansion of the oceans themselves (Hugonnet et al., 2021).

'What Anglo is doing to our glaciers, they're doing to all of humanity,' says Uranga.

Opposing companies as rich and powerful as Anglo American is never an easy undertaking, but in Chile it is particularly difficult given the country's long history of mining. It is sometimes said that Chile is *'un país minero'* – a mining country. For environmental activists and people in mining-affected communities, the challenge is to reshape these popular and deeply ingrained beliefs.

'What we're trying to build is the idea that Chile is a glacial country, *not* a mining country,' says Uranga. 'That this is what our wealth consists of.'

There are some signs that perceptions are changing. Constanza San Juan Standen is an activist with the Coordination of Territories in Defence of the Glaciers, a network of groups all over Chile which is campaigning for a proper glaciers law.

'In the past if you said that you were against mining everyone thought you were mad, but now in certain circles people are starting to recognize that mining isn't the future,' she says.

'A transversal issue'

Chile's water crisis is above all a question of social justice. The massive concentration of water rights is a consequence of the radical neoliberal economic model imposed at the barrel of a gun by the Pinochet dictatorship, and left largely intact by the democratic governments that followed.

While this model produced a period of sustained economic growth higher than the average for Latin America, this meant little to most Chileans, who have struggled with insecure employment, low wages, and a relatively high cost of living. Meanwhile, the Chilean elite has become immensely richer, in many cases profiting from extractive industries such as mining, agribusiness, forestry, and fish farming.

The consequences of this model are now plain to see, with the environmental impacts of these industries coinciding with an unprecedented period of drought. The result is a country which cannot guarantee a reliable water supply to many of its citizens, despite having amongst the highest figures for GDP per capita in Latin America.

'The whole economic model is built to always benefit the same economic groups while leaving most of us completely impoverished, and it's a total impoverishment: not only do we not have environmental justice, nor do we have social justice. These two things are inextricably linked – one can't exist without the other,' says Stefanía Vega, another activist with the Coordination of Territories in Defence of the Glaciers.

'I think the government is playing with fire because it's going to come to a point – and by no means is this what I want – where people are going to erupt with fury and riot because there's such a lack of respect,' San Juan told me in 2016.

This is precisely what happened in 2019, in Chile's *estallido social* [literally, the 'social outburst'], the biggest wave of civil unrest since the return to democracy, with millions of protestors taking to the streets in cities all over the country.

While most of the protests were peaceful, some were violently repressed. Thirty people died, while more than 8,000 were injured, including more than 400 cases of eye trauma, mostly from rubber bullets and tear gas canisters fired by the Carabineros, the Chilean police. There was also excessive use

Photo 6.4 View of the *estallido social* in Santiago / Hugo Morales 2019 / CC BY-SA 4.0

of preventive detention, and hundreds of detainees were subject to sexual violence and torture.

A report by Amnesty International stressed that these were not isolated incidents, but part of a calculated strategy of excessive force by the Carabineros, with the responsibility for the abuses going all the way to the top of the institution. Though Amnesty did not investigate the role of other actors beyond the Carabineros, it did criticize Sebastián Piñera's government for failing to control them and urged that those responsible be held accountable (Amnesty International, 2020, pp. 6–7).

While much of the analysis of the *estallido social* tended to focus on issues such as public transport, health, education, and the high cost of living, environmental issues and particularly access to water were also factors.

'When the *estallido social* happened in 2019, we started having these public meetings where everyone got together and discussed what was bothering them,' says Ximena Gallardo. 'And what we realized was that water was a transversal issue affecting everyone' – in other words, it cut across all the other areas of grievance felt by Chileans.

In response to the unrest, the Chilean Congress agreed to hold a referendum in 2020 on whether to scrap the dictatorship-era Constitution of 1980, and, if so, how a new constitution would be written. The vote, which was eventually held in October 2020, was decisive, with over 78 per cent voting in favour of a new constitution. Significantly, 79 per cent also voted in favour of the document being drafted by a fully elected constitutional assembly. In May 2021, these elections were held, with voters largely rejecting the traditional parties of the right and centre-left which have dominated Chilean politics since the return to democracy.

The big winners were the independent candidates: 88 of the 155 people elected to the assembly were independents (64 per cent), many of whom were associated with local social, territorial, and environmental movements. One of them was Constanza San Juan, elected to represent the Atacama Region in the north of Chile, one of at least 57 delegates who signalled an interest in overhauling Chile's system of water management to prioritize human consumption over industrial use (Fundación Terram, 2021). Indeed, the eventual draft of the new constitution stated that 'The exercise of the human right to water, sanitation, and ecological equilibrium, will always take priority' (Convención Constitucional, 2022, p. 103).

Unfortunately for the country's mining-affected communities, the Chilean people rejected the draft in a referendum in September 2022 – a result which surprised many. Still, all is not lost. While the draft was certainly overlong and probably overambitious, its rejection is not a call for the dictatorship-era Constitution to remain in place. A successful slogan deployed by the No campaign was 'Así No' – 'Not like this' – rather than a blanket rejection of the initiative per se (Gregg, 2022).

In December 2022, Chile's President Gabriel Boric announced that an agreement had been reached on writing a new draft, this time with a much stronger role for Chile's two houses of Congress and with more power for the established political parties (Molina, 2022). Nonetheless, the need to reform the existing system of water management remains as urgent as ever. It will continue to feature in debates on the new constitution, as well as on the wider political agenda.

III. The foundation of the resistance

Elsewhere in Latin America, in regions where local communities have mounted successful opposition to mining projects, arguments around the defence of water have almost always been key.

One good example comes from the Argentine province of Mendoza. Located just 180 km from Santiago on the other side of the *cordillera*, Mendoza has a law (7722), which forbids the use of toxic substances such as cyanide, mercury, sulphuric acid, and others in mining processes. The aim is to protect the province's natural resources and particularly its water.

Mendoza is in a rain shadow: the Andes block most precipitation from falling in the province, meaning it enjoys more than 300 days of sunshine a year and only 200 mm of rainfall. And yet, one of the first things that any visitor to the eponymous provincial capital will notice is the abundant vegetation; leafy, tree-lined streets; lush, shady squares and parks.

This is only possible thanks to an ingenious system of irrigation with its roots in the pre-Columbian era. Glaciers and snowmelt on the high slopes of the Andes feed Mendoza's rivers, which, in turn, supply a network of canals crisscrossing the provincial capital, guaranteeing water for its public spaces. Water is then directed towards the agricultural areas of the province (Bormida, 1984, p. 129). This infrastructure has long been accompanied by robust environmental regulation aimed at securing a sustainable water supply.

'Mendoza has had a water law since the end of the 19th century. The first Argentines to attend an international conference on the environment were from Mendoza,' says Lucrecia Wagner, a researcher based at the National Scientific and Technical Research Council (CONICET) in Mendoza. 'They were the godfathers of the water legislation, which later came to influence other environmental law.'

The natural scarcity of water means that many *mendocinos* are very conscious of the need to manage resources carefully and suspicious of any enterprise which may put them in danger. Law 7722 was obtained thanks to popular pressure from community assemblies and social movements and these groups continue to defend the law whenever an attempt is made to overturn it, as in late 2019, when the provincial government modified the law to permit the use of cyanide and sulphuric acid. The strength of popular opposition was such that the government was forced into a humiliating U-turn just ten days later.

'The province has generated great hope,' says Wagner. 'It's an example of what an organized society can achieve when it stands up for what it believes to be fair.' (Navicelli, 2020)

More recently, in February 2021, in Cuenca, Ecuador, a *consulta popular* – a kind of local referendum – was held on whether to ban mining in the watersheds of five rivers in the area. Cuenca is close to two protected areas, home to *páramos* (unique high-altitude wetlands found only in the northern Andes and in a few small areas in Central America), lagoons and cloud forest, as well as unique species of flora and fauna. The city depends on these highland areas for its water and the campaign for the mining ban, led by the People's Council for the Water of Cuenca, focused on the importance of water not just for drinking, but for local culture, traditions, and enterprise – such as agriculture, pottery, even baptism. It was successful: 80 per cent voted in favour of the ban.

'The results of the *consulta popular* represent a turn of the screw in the process of defending our water,' says David Fajardo Torres, who worked for the People's Council on the referendum campaign. 'Now we have legitimacy, in that practically eight out of ten people in Cuenca are against mining and don't want the development of these projects.'

'And this is a sovereign decision,' he adds. 'A decision, moreover, taken via a direct democracy mechanism which can't easily be ignored by the central government or any other authority. So the political momentum is in our favour.'

***

'Water is worth more than gold.' 'We can live without gold; we die without water.' 'You are 75 per cent water and 0 per cent gold. Which do you defend?'

These are just some of the various slogans used to great effect by environmentalists and anti-mining activists all over Latin America. The defence of water has proven to be a powerful rallying cry in mobilizing opposition to the mining industry, for two principal reasons.

Firstly, the impacts of large mining operations on both the quantity and quality of water are undeniable. Mining-affected communities all over the region consistently relate the same stories: wherever operations are located, the industry tends to significantly increase pressure on water resources, leaving less for domestic consumption and local enterprise, particularly agriculture. Meanwhile, contamination of aquifers and other water sources forces communities either to obtain water from elsewhere – sometimes at great cost and inconvenience – or risk illness, perhaps even death.

Secondly, the simplicity of the message ensures its uptake. In the remote, sometimes poor communities where mining companies usually set up, people may not be receptive to arguments around sustainability, climate change, or the need to safeguard biodiversity. However, if they believe that the water they depend on for drinking, cooking, and washing is in danger, it is a different matter.

As Wagner says, 'The defence of water is the foundation for the rejection of *megaminería*' (Navicelli, 2020). This is clear in the significant lengths that most companies go to in order to address concerns about potential impacts on water supply. Water is their Achilles heel – and they know it only too well.

Notes

1. Signed in 1494, before the first Portuguese explorers had even landed in Brazil, the treaty drew a vertical line through the eastern half of the country. Everything on the eastern side was Portuguese; everything on the western side, Spanish.
2. Exchange rates used were those given at time of publication of source.

References

All references to web-based material were checked and still available in November 2022 unless otherwise stated.

All references are listed, with clickable links for your convenience, on the page for this chapter on the Heart of Our Earth website: <https://lab.org.uk/the-heart-of-our-earth/>

Amazon Watch (2021) 'Mining Company Anglo American States It Does Not Rule Out Mining on Indigenous Lands in the Brazilian Amazon'. [online] Available at: <https://amazonwatch.org/news/2021/0209-mining-company-anglo-american-states-it-does-not-rule-out-mining-on-indigenous-lands-in-the-brazilian-amazon>.

Ameonna, E. (2020) '$222 mil millones gastó el Estado en contratar camiones aljibe y más de $11 mil millones en comprar forraje para ganado'. [online]

Vergara 240. Available at: <https://vergara240.udp.cl/especiales/222-mil-millones-gasto-el-estado-en-contratar-camiones-aljibe-y-mas-de-11-mil-millones-en-comprar-forraje-para-ganado/>.

Amnesty International (2020) *Ojos sobre Chile: violencia policial y responsabi-lidad de mando durante el estallido social.* [online] Amnesty International. Available at: <https://www.amnesty.org/es/latest/research/2020/10/eyes-on-chile-police-violence-at-protests/>.

APIB and Amazon Watch (2021) *Complicity in Destruction IV: how mining companies and international investors drive indigenous rights violations and threaten the future of the Amazon.* [online] Amazon Watch. Available at: <https://amazonwatch.org/assets/files/2022-complicity-in-destruction-iv.pdf>.

Aragão, T. (2021) 'Garimpo na Terra Indígena Munduruku cresce 363% em 2 anos, aponta levantamento do ISA'. [online] Instituto Socioambiental. Available at: <https://www.socioambiental.org/en/node/7277>.

Arellano, A. (2020) 'Mercado del agua: el rentable negocio de un empresario que vendió en $184 millones al Estado derechos de agua en Petorca'. [online] Vergara 240. Available at: <https://vergara240.udp.cl/especiales/sequia-en-chile-mercado-del-agua-petorca-v-region/#:~:text=ESPECIALES%20 V240-,Mercado%20del%20agua%3A%20el%20rentable%20negocio%20 de%20un%20empresario%20que,un%20total%20de%20%24303%20 millones>.

Audi, A. (2018) 'O passado garimpeiro de Bolsonaro'. [online] The Intercept Brasil. Available at: <https://theintercept.com/2018/11/05/passado-garim peiro-bolsonaro/>.

Bormida, E. (1984) 'Mendoza, una ciudad oasis'. *Revista de la Universidad de Mendoza*, [online] 4/5, pp. 121–137. Available at: <http://www.um.edu. ar/ojs2019/index.php/RUM/article/view/189>.

Brazil Matters (2021) *Indigenous emergency – The Fight for Life in Brazil* [webinar]. 29 April 2021.

Brenning, A. and Azócar, G. (2010) 'Minería y glaciares rocosos: impactos ambientales, antecedentes políticos y legales, y perspectivas futuras'. *Revista de Geografía Norte Grande*, [online] 47, pp. 143–158. Available at: <https:// www.scielo.cl/scielo.php?pid=S0718-34022010000300008&script=sci_ arttext#c2>.

Brown, K. (2012) *A History of Mining in Latin America. From the Colonial Era to the Present.* Albuquerque: University of New Mexico Press.

Carneiro, T. (2020) 'MPF questiona FAB sobre voo que levou garimpeiros do PA para reunião com ministro Salles em Brasília'. [online] G1. Available at: <https://g1.globo.com/pa/para/noticia/2020/09/28/mpf-questiona-fab-sobre-voo-que-levou-garimpeiros-do-pa-para-reuniao-com-ministro-salles-em-brasilia.ghtml>.

CIVICUS (2020) "Chile has entirely privatised water, which means that theft is institutionalised". [online] Available at: <https://www.civicus. org/index.php/media-resources/news/interviews/4271-chile-has-entirely-privatised-water-which-means-that-theft-is-institutionalised>.

Comissão da Verdade do Estado de São Paulo (2015) 'Violação aos direitos dos povos indígenas'. [online] Available at: <http://comissaodaverdade.al.sp. gov.br/relatorio/tomo-i/parte-ii-cap2.html>.

Console, L. (2019) 'Tapajós tóxico: garimpo aumenta níveis de mercúrio no rio e população adoece'. [online] *Brasil de Fato*. Available at: <https://www.brasildefato.com.br/2019/03/16/tapajos-toxico-garimpo-aumenta-niveis-de-mercurio-no-rio-e-populacao-adoece>.

Convención Constitucional (2022) *Borrador Nueva Constitución*. [online] Available at: <https://www.chileconvencion.cl/wp-content/uploads/2022/05/PROPUESTA-DE-BORRADOR-CONSTITUCIONAL-14.05.22-1-1.pdf>.

Crespo-Lopez, M., Augusto-Oliveira, M., Lopes-Araújo, A., Santos-Sacramento, L., Yuki Takeda, P., de Matos Macchi, B., Martins do Nascimento, J., Maia, C., Lima, R. and Arrifano, G. (2021) 'Mercury: What can we learn from the Amazon?'. *Environment International*, [online] 146. Available at: <https://doi.org/10.1016/j.envint.2020.106223>.

Dias, J., Cavalcanti Filho, J., Kehl, M., Pinheiro, P., de Abreu Dallari, P. and Cardoso de Cunha, R. (2014) *Comissão Nacional da Verdade. Relatório. Volume II. Textos Temáticos*. [online] Comissão Nacional da Verdade. Available at: <http://cnv.memoriasreveladas.gov.br/images/pdf/relatorio/volume_2_digital.pdf>.

Dobry, R. (1965) 'Efectos del cismo de marzo de 1965 en los tranques de relaves de El Cobre'. *Revista del Idiem*, [online] 4(2), pp. 85-107. Available at: <https://core.ac.uk/download/pdf/132237474.pdf>.

El Desconcierto (2020) 'Desigualdad hídrica: Un 1% posee el 79% del volumen total de agua disponible'. [online] Available at: <https://www.eldesconcierto.cl/sociedad-colaborativa/2020/12/07/desigualdad-hidrica-un-1-posee-el-79-del-volumen-total-de-agua-disponible.html>.

El Observador (2020) 'El verdadero estado del agua en El Melón: ¿Es cierto que la minería tiene seco a este distrito de Nogales?'. [online] pp. 32–33. Available at: <https://www.litoralpress.cl/sitio/Prensa_Texto?LPKey=x.J.B95.Ff9.O.Ü1x.C.E1kyskmjyjz.N0en7yixhldau8.G.Hws.I.Ö>.

Emmet Hernan, R. (2022) 'Minamata, Japan'. [online] Irish Environment. Available at: <https://www.irishenvironment.com/blog/minamata-japan-1950s/>.

Fundación Terram (2018) 'Comunidad se opone a Angloamerican y su Tranque de Relaves El Torito'. [online] Available at: <https://www.terram.cl/2018/01/comunidad-se-opone-a-angloamerican-y-su-tranque-de-relaves-el-torito/>.

Fundación Terram (2021) 'Constituyentes por el Agua presentan iniciativa para garantizar el Derecho Humano al agua y Saneamiento en la Nueva Constitución'. [online] Available at: <https://www.terram.cl/2021/12/constituyentes-por-el-agua-presentan-iniciativa-para-garantizar-el-derecho-humano-al-agua-y-saneamiento-en-la-nueva-constitucion/>.

Gobierno de Chile (2019) 'Día Mundial Del Agua: Ministro Moreno Destacó El Trabajo De Compromiso País En Favor De 1 Millón 400 Mil Personas Sin Acceso Al Agua Potable'. [online] Available at: <https://www.gob.cl/noticias/dia-mundial-del-agua-ministro-moreno-destaco-el-trabajo-de-compromiso-pais-en-favor-de-1-millon-400-mil-personas-sin-acceso-al-agua-potable/>.

Greenpeace Chile (2020) 'El Melón y su batalla por no secarse: Paltas y minería, los poderosos actores que dejan sin agua la zona central de Chile.'. [online] Available at: <https://www.greenpeace.org/chile/noticia/uncategorized/

el-melon-y-su-batalla-por-no-secarse-paltas-y-mineria-los-poderosos-actores-que-dejan-sin-agua-la-zona-central-de-chile/>.

Gregg, E. (2022) 'Chile: why was the new constitution rejected?' [online] Latin America Bureau. Available at: <https://lab.org.uk/chile-why-was-the-new-constitution-rejected/>.

Guzmán, L. (2019) 'Chile's mega-drought rolls on'. [online] Diálogo Chino. Available at: <https://dialogochino.net/en/climate-energy/30820-chiles-mega-drought-rolls-on/>.

Hugonnet, R., McNabb, R., Berthier, E. et al. (2021) 'Accelerated global glacier mass loss in the early twenty-first century'. *Nature*, [online] 592, pp. 726–731. Available at: <https://doi.org/10.1038/s41586-021-03436-z>.

Jamasmie, C. (2019) 'Anglo vows to quit Los Bronces $3B expansion if it affects nearby glaciers'. [online] Mining.com. Available at: <https://www.mining.com/anglo-vows-quit-los-bronces-3b-expansion-affects-nearby-glaciers/>.

Jordan, L. (2021) 'Indigenous 'blood gold' almost certainly exported to UK, say Brazilian researchers'. [online] Unearthed. Available at: <https://unearthed.greenpeace.org/2021/09/01/amazon-blood-gold-likely-being-imported-by-uk-researchers-say/>.

Jordan, L. and Potter, H. (2021) 'Anglo American, eyeing copper on protected indigenous land, failed to consult Amazonian community'. [online] InfoAmazonia. Available at: <https://infoamazonia.org/en/2021/04/15/anglo-american-eyeing-copper-on-protected-indigenous-land-failed-to-consult-amazonian-community/>

Kotscho, R. (2021) 'Serra Pelada fluvial: do major Curió ao capitão Bolsonaro, a saga da morte'. [online] UOL. Available at: <https://noticias.uol.com.br/colunas/balaio-do-kotscho/2021/11/25/serra-pelada-fluvial-do-major-curio-ao-capitao-bolsonaro-a-saga-da-morte.htm>.

Kuyek, J. (2019) *Unearthing Justice. How to protect your community from the mining industry*. Toronto: Between The Lines.

London Mining Network (2020) 'Questions for Anglo American'. [online] Available at: <https://londonminingnetwork.org/2020/05/questions-for-anglo-american/>.

Lopes, M. (2018) 'Exploração de ouro no Brasil começou em São Paulo - e a região pode conter pepitas até hoje, dizem especialistas'. [online] BBC News Brasil. Available at: <https://www.bbc.com/portuguese/brasil-44988574>.

Marconi, C. and Urquiza, N. (2020) 'Vecinos de El Melón se enfrentan a Anglo American por el agua'. [online] Vergara 240. Available at: <https://vergara240.udp.cl/sequia-lucha-por-el-agua-habitantes-el-melon-con-anglo-american/>.

Millan Lombrana, L., Campano, M. and Dodge, S. (2019) 'South America's glaciers may have a bigger problem than climate change'. [online] Mining.com. Available at: <https://www.mining.com/web/south-americas-glaciers-may-have-a-bigger-problem-than-climate-change/>.

Minai, M. (2018) 'Methylmercury and Human Embryonic Development'. [online] The Embryo Project Encyclopedia. Available at: <https://embryo.asu.edu/pages/methylmercury-and-human-embryonic-development>.

Minas Júnior Consultoria Mineral (2019) 'Garimpo, uma atividade ilegal?'. [online] Available at: <https://www.minasjr.com.br/garimpo-uma-atividade-ilegal/>.

Miranda, R. (2021a) 'The Munduruku Demand That Mining Giant Anglo American Publicly Commit to Not Mine on Indigenous Lands'. [online] Amazon Watch. Available at: <https://amazonwatch.org/news/2021/0108-munduruku-demand-that-anglo-american-publicly-commit-to-not-mine-on-indigenous-lands>.

Miranda, R. (2021b) 'Munduruku Victory: Anglo American Agrees to Withdraw 27 Mining Research Permits in Territories Following Sustained Pressure by Indigenous Movement'. [online] Amazon Watch. Available at: <https://amazonwatch.org/news/2021/0721-munduruku-victory-sustained-pressure-forces-anglo-american-to-withdraw-mining-permits>.

Molina, P. (2022) '3 preguntas para entender cómo se escribirá la nueva Constitución de Chile tras el apabullante rechazo en el proceso anterior'. [online] BBC News Mundo. Available at: <https://www.bbc.com/mundo/noticias-america-latina-63965925>.

Movimiento por el Agua y los Territorios – MAT (2019) *Capitulo_01_Código de Aguas*. [video] Available at: <https://www.youtube.com/watch?v=0v7F9IKfTJc&t=568s>.

Navicelli, V. (2020) 'Mujeres que marcaron el rumbo del agua en Mendoza'. [online] *Los Andes*. Available at: <https://www.losandes.com.ar/conoce-los-senderos-del-agua-mendocina/>.

Nikou, E. and Moncau, J. (2021a) 'Apoio de Bolsonaro a garimpo coloca em risco a vida de duas líderes Munduruku'. [online] Repórter Brasil. Available at: <https://reporterbrasil.org.br/2021/02/apoio-de-bolsonaro-a-garimpo-coloca-em-risco-a-vida-de-duas-lideres-munduruku/>.

Nikou, E. and Moncau, J. (2021b) 'As mulheres Munduruku estão envenenadas por mercúrio e temos provas', denuncia líder indígena'. [online] Repórter Brasil. Available at: <https://reporterbrasil.org.br/2021/02/as-mulheres-munduruku-estao-envenenadas-por-mercurio-e-temos-provas-denuncia-lider-indigena/>.

Portal Minero (2014) 'SMA multa a Anglo American Sur por incumplimientos ambientales en El Soldado'. [online] Available at: <http://www.portalminero.com/display/NOT/2014/09/02/SMA+multa+a+Anglo+American+Sur+por+incumplimientos+ambientales+en+El+Soldado>.

Prosser, I., Wolf, L. and Littleboy, A. (2011) 'Water in mining and industry'. In: I. Prosser, ed., *Science and Solutions for Australia. Water.* [online] Collingwood, Victoria: CSIRO Publishing, pp. 135–146. Available at: <https://www.publish.csiro.au/ebook/download/pdf/6557>.

Santibáñez, L., Roblero, S. and Giordano, G. (2020) '¿Qué pasa con el agua en El Melón?'. [online] El Mostrador. Available at: <https://www.elmostrador.cl/agenda-pais/2020/09/29/que-pasa-con-el-agua-en-el-melon/>.

Surf Beats Radio (2020) 'Los Bronces Integrado: La polémica expansión de Anglo American que dejaría sin agua a Santiago y destruirá importantes glaciares'. [online] Available at: <https://surfbeatsradio.com/los-bronces-integrado-la-polemica-expansion-de-anglo-american-que-dejaria-sin-agua-a-santiago-y-destruira-importantes-glaciares/>.

Voiland, A. (2019) 'Long-Term Drought Parches Chile'. [online] NASA Earth Observatory. Available at: <https://earthobservatory.nasa.gov/images/145874/long-term-drought-parches-chile>.

Voiland, A. (2020) 'A Strained Water System in Chile'. [online] NASA Earth Observatory. Available at: <https://earthobservatory.nasa.gov/images/146577/a-strained-water-system-in-chile>.

Volckhausen, T. (2020) 'Niobium mining in Brazilian Amazon would cause significant forest loss: Study'. [online] Mongabay. Available at: <https://news.mongabay.com/2020/07/niobium-mining-in-brazilian-amazon-would-cause-significant-forest-loss-study/>.

World Health Organization (2020) '10 chemicals of public health concern'. [online] Available at: <https://www.who.int/news-room/photo-story/photo-story-detail/10-chemicals-of-public-health-concern>.

CHAPTER 7
The law of holes: Mining and climate change

If humanity is to prevent the worst impacts of climate change, it is essential that we decarbonize our economies. This will involve the widespread rollout of technologies which are enormously metal- and mineral-intensive. But mining is not without its social and environmental impacts, nor is it free of responsibility for the climate emergency. This chapter assesses to what extent mining may be part of the solution, looking at two metals which will be essential for the energy transition: copper, focusing on the Cordillera del Cóndor, Ecuador; and lithium, focusing on the Salar de Atacama in Chile.

Humanity is in a major hole. The summer of 2021 brought catastrophic flooding which killed more than 100 people in Germany and Belgium; a blistering heatwave in the western United States and Canada which pushed temperatures close to 50 °C and caused hundreds of deaths; plus raging wildfires across 13 U.S. states, the Mediterranean, and even Siberia.

The ferocity of these hitherto freak weather events has taken even scientists by surprise. We must face facts: climate change is a reality, and our failure to heed repeated warnings from climatologists since at least the 1970s now threatens our climate system with severe and even irreversible damage, with consequences far more serious than those seen in 2021.

According to the Climate Action Tracker, the world is heading for an increase of around 2.7 °C by 2100. Even their 'optimistic' analysis predicts warming of 2.1 °C, in excess of the limit set by the Paris Agreement of 'well below two degrees Celsius above pre-industrial levels' (McGrath, 2020). Even if – and it's a big 'if' – global warming is kept below 2 °C, scientists warn that the impacts across much of the world will still be devastating – and will get worse with each fraction of a degree increase (Roberts, 2020).

Is this a hole we can, quite literally, dig ourselves out of? For the mining industry, the answer is yes.

'Mining has a critical role to play in providing the metals and minerals needed for a low-carbon world,' states Anglo American (Anglo American, 2021). 'We have an important role to play in supporting and enabling the transition to net zero emissions,' echoes Rio Tinto (Rio Tinto, 2021).

Indeed, the transition away from fossil fuels towards so-called 'green technologies' will be enormously metal- and mineral-intensive. A World Bank report from 2020 estimated that production of graphite, lithium, and cobalt may have to increase by as much as 500 per cent by 2050 to meet growing demand (World Bank, 2020). These metals and minerals, and others, are required for renewable energy technologies such as solar panels and wind turbines; they are needed for energy storage, to ensure a constant power supply; and they are an essential component of the batteries which power electric vehicles (EVs).

This chapter will describe the implications for communities and the environment in the case of two of these transition metals, namely copper and lithium, reporting first on the expansion of copper mining into the Cordillera del Cóndor in Ecuador, and then on the burgeoning lithium industry in Chile's Atacama Desert.

I. The Cordillera del Cóndor, Ecuador: mining the wilderness

'They destroyed my home with bulldozers in the early hours of the morning,' says Luis Sanchez Zhiminaycela. 'We lost everything for having resisted.'

Sanchez, an indigenous Cañari-Kichwa, is vice-president of CASCOMI, a coalition of local and regional campesinos, indigenous communities, and environmentalists, which has coalesced opposition to mining in the Cordillera del Cóndor, a mountain range which straddles the border with Peru in the far south-east of Ecuador (Sánchez-Vázquez and Leifsen, 2019).

Sanchez's former home stood in the village of San Marcos, in the parish of Tundayme. Residents from at least 32 families are thought to have been displaced – over a third of whom were children – on three occasions between September 2015 and February 2016. They were given no warning. Police and private security guards arrived between the hours of 4 a.m. and 6 a.m. and informed them they had five minutes to leave their property. Those who tried to resist were threatened, intimidated, and in some cases, beaten. Their properties were destroyed, the debris buried. Some of those evicted were left with nothing but the clothes they were wearing (Morán, 2019).

Today, nothing is left of San Marcos. It was cleared to make way for a tailings dam, one of two belonging to Mirador, an open-pit copper mine run by EcuaCorriente S.A, a joint venture between the state-owned Chinese companies Tongling Nonferrous Metals Group and the China Railway Construction Corporation (CRCC). It is the largest mine in Ecuadorian history and the country's first ever open-pit operation, projected to produce 3.18 million tonnes of copper – plus smaller quantities of gold and silver – during a thirty-year lifespan (Pérez, 2019).

Unlike its neighbour Peru, and Chile further to the south, Ecuador has no history of industrial mining. While it has been a major oil exporter since the 1970s, in recent decades successive governments have sought to diversify and capitalize on the country's rich, largely untapped mineral resources. Not only does Ecuador have significant reserves of gold and silver, it also lies on the Andean Copper Belt – a 2,500-kilometre-long arc

Photo 7.1 Luis Sanchez points towards the Mirador site / © Andrés Bermúdez Liévano/ Diálogo Chino 2019 CC BY-NC-ND 4.0

rich in copper running through the Andes from Chile as far as Colombia. The area of south-east Ecuador through which the belt runs is thought to be one of the few undeveloped copper regions left on Earth (Corriente Resources Inc., n.d.).

'It's very likely that Ecuador has 13 or 14 per cent of global copper reserves,' said the then president, Lenín Moreno, at the World Economic Forum in 2019 (Fundación Pachamama, 2020).

This is an exaggeration, but it's indicative of the haste with which Ecuador has been courting the global mining industry. Fifteen per cent of national territory is currently under concession, with these concessions often overlapping with areas of great ecological sensitivity: *páramos* (high-altitude wetlands), cloud forests, and the headwaters of streams and rivers that flow into the Amazon region (ibid.).

Mirador is a case in point. The Cordillera del Cóndor is where the Andes meet the Amazon, a wild, mysterious landscape of plunging cliffs and valleys; strange tepui formations (flat-topped elevations) like those found in the jungles of Venezuela and Guyana; and mountains covered in dense, dark vegetation, their peaks veiled in cloud. It is thought to be one of the most biodiverse areas on the planet.

'The problem is that this is a mega-diverse territory and it has many water sources,' says Sanchez. 'The streams and rivers that begin in the Cordillera flow into the Río Zamora, and in turn the Río Santiago, the Río Marañón, and of course the Amazon.'

Mirador: 'a national strategic project'

The origins of mining in the Cordillera del Cóndor go back to 1995, when Ecuador and Peru fought a brief war over an area on the Peruvian side of the Cordillera which Ecuador had long claimed as its own, in a territorial dispute that dates from the founding of the two countries.

In 1998, a peace treaty was signed in Brasília, in which Ecuador renounced its claim on the disputed territory and both countries promised to create 'peace parks': protected areas on both sides of the border which would help to safeguard the region's rich biodiversity and prevent future conflict. While Peru created a 152,000-hectare national park, Ecuador opted to protect four, much smaller areas, amounting to just 41,000 hectares (Pérez, 2019). Around the same time, the first mining companies began arriving in the area.

'After the peace treaty was signed with Peru, the mining industry entered with the blessing of the government at the time,' explains Sanchez. 'Then in '99 there was an economic crisis, which led to the change in currency to the dollar.[1] This served as a pretext for ushering in mining, supposedly as an alternative source of employment for people.'

Corriente Resources, a Canadian company, began exploration work in the year 2000, attempting to persuade locals to sell their land by promising jobs and carrying out some modest development projects (Gordon and Webber, 2016, p. 218). But the strategy largely failed to convince the indigenous Shuar, whose ancestral territory encompasses the Cordillera del Cóndor.

The largest indigenous group in the Ecuadorian Amazon, the Shuar are famous for having successfully resisted colonization, both by the Incas and the Spanish. In more recent times, they have a history of highly successful organization and education, conflict with oil multinationals, and later with miners (ibid.).

In November 2006, the Shuar set about evicting workers from mining camps, aiming to 'cleanse' their territory of mining. This was followed in December by a three-day confrontation at the Mirador camp between mining opponents and police and private security (ibid., p. 219). Given the ferocity of the opposition, the government was forced to suspend Corriente's activity in both the provinces of Zamora Chinchipe – in which Mirador is located – and neighbouring Morona Santiago to the north, though the suspension was never fully implemented (Warnaars, 2013, p. 156).

But in 2010 Corriente Resources was taken over by the CRCC-Tongling consortium, and in the ensuing years company strategy became increasingly aggressive.

'At first [under Corriente Resources] they tried to obtain the land through deception, trying to buy people off with false promises,' says Sanchez. 'But people resisted, they didn't want to leave their land. So they had to try other strategies based on the mining law, like mining easement, so they could use the police and military to evict people.'

Mining easement is a mechanism which allows the state to order the rental of a property for up to three decades if it considers the area to be in the national interest (Bermúdez Liévano, 2019). In this case, it was tantamount to an eviction order: the compensation offered to families was not enough to cover their losses and those who refused it were still expelled from their homes (Colectivo de Investigación y Acción Psicosocial, 2017, p. 70).

The evictions did not stop in 2016. In April 2022, an elderly couple in Tundayme were informed that their property fell within EcuaCorriente's concessions; on the 26 of that month, their house was dismantled by a group of 25 police officers. There is concern that this will be just the first in a new wave of evictions in the area (Alvorado, 2022).

EcuaCorriente has denied responsibility for the evictions, claiming it attempted to negotiate the purchase of the land in good faith. The company says the decision to evict families was taken by the Ecuadorian state.

'The mine is a national strategic project,' says Zhu Jun, community relations manager for EcuaCorriente. 'If agreement can't be reached with the locals, the government can assert its rights out of strategic need. It wasn't our company that asserted those rights. It was the government's solution, implemented entirely by the government.' (Hui, 2019)

This is contested by those evicted from San Marcos, whose testimonies refer not only to the presence of the police and other state officials during the evictions, but also to private security guards employed by the company (Colectivo de Investigación y Acción Psicosocial, 2017, p. 234). Several of the women evicted complained of sexual harassment by a company employee known as 'Negro S'.

'One man from the company, who goes by the name of 'Negro S', offered to take us to a house in Gualaquiza,' reads a testimony by one woman. 'He harassed me, my sister, my aunt and my mum. And it wasn't the first time.' (ibid., p. 80)

A history of dispossession

CASCOMI has brought several legal challenges against the Ecuadorian state and the company since the evictions. It has been to no avail, however, and the community is running out of options.

The courts have exploited a history of internal migration to the Cordillera as a means of denying communities the right to consultation. Though the area lies within the ancestral territory of the Shuar, in the mid-20th century settlers from elsewhere in the country began arriving, particularly from the southern highlands.

'We Kichwa, Cañari-Kichwa, are from the provinces of Cañar and Azuay, in the southern Andes of Ecuador,' says Sanchez. 'Our grandparents have been here since the 1950s. We were born here and we live here still. So there's us Kichwa, the Shuar, and some who identify as campesinos.'

The courts argue, therefore, that Tundayme is not an ancestral indigenous community and those within the parish who were displaced were not entitled to consultation, as required by both the Ecuadorian Constitution and Convention 169 of the International Labour Organization (ILO).

'I'm not in any doubt,' said the judge Carlos Dávila, when rejecting a challenge brought by CASCOMI against EcuaCorriente in 2019. 'This is not an ancestral indigenous community; it is the result of migration from the south. It's not a community to which consultation applies.' (Morán, 2019)

Of course, a community's indigeneity should have no bearing on whether residents are evicted from their homes at a moment's notice. But the loss is particularly bitter in the case of San Marcos, given that many of those who were displaced *do* identify as indigenous.

'The people who live there, the Shuar and us Kichwa, we know who we are and where we come from ... To say that we're not an ancestral community is just a way of discriminating against our people,' says Sanchez (ibid.).

Migration continued during the 1960s, when, aiming to colonize a remote corner of the country and create a 'live border' with Peru, the military government at the time began offering land titles to settlers over the heads of the Shuar. Indigenous people were sometimes cheated out of their lands, signing agreements under the effects of alcohol (Warnaars, 2013, p. 158). Over the decades, these land titles have been bought and sold, with some now in the hands of transnational mining companies. While for the Shuar the territory is ancestral, as far as the companies are concerned, they bought the land from its legitimate owners – and they have the papers to prove it.

'The company owns that land, but that transaction was not really legitimate. If you go backwards, you're going to find there was an adjudication by some state organization. That's the beginning of the dispossession of the Shuar,' says Verónica Potes, a lawyer specializing in indigenous rights (Ponce Ycaza, 2019).

The ethnic and cultural mix of these communities also means that they do not necessarily have a common position when it comes to mining, something which works in favour of the companies. That said, the dynamics of resistance are complex. The Shuar are not uniformly anti-mining, just as not all the campesinos – mostly the descendants of migrants – are in favour. However, given the Shuar's history of militancy, in general, the companies have found it harder to advance in Shuar areas.

'In the Tundayme area, where Mirador is located, there is a Shuar population ... but there is greater presence of migrants, people from other parts of the country,' explains Andrés Tapia, director of communications at the Confederation of Indigenous Nationalities of the Ecuadorian Amazon (CONFENIAE). 'These campesino communities aren't as organized, and it's much easier for companies to break communities where there's no tradition of struggle and resistance and where the presence of the Shuar isn't as strong. This is why Mirador has managed to advance, though of course there has been resistance.'

San Carlos-Panantza: the Shuar vs the state

In March 2020, just 40 km to the north of Mirador, in the neighbouring province of Morona Santiago, persons unknown broke into the mining camp La Esperanza, where they set fire to the installations and made off with equipment and other items of value.

La Esperanza is part of the mining complex San Carlos-Panantza. Like Mirador, it is a copper project, and again, its owners are Tongling and CRCC, via a subsidiary called ExplorCobres S.A. Currently, the project is stuck at the advanced exploration stage. If it manages to progress, it will be an even larger undertaking than Mirador. Copper reserves are estimated at 6.6 million tonnes – more than double those of its sister project to the south (Ruiz Leotaud, 2020).

But this latest act of sabotage serves as a reminder that one of Ecuador's most volatile mining conflicts remains live and could very easily explode into violence once more.

In August 2016, hundreds of armed police were deployed to evict a community known as Nankints, home to 32 Shuar. The villagers were told that the land belonged to ExplorCobres and that they were trespassing. Like the residents of San Marcos before them, they were given just minutes to grab whatever they could before their property was razed (Ponce Ycaza, 2019). The ground on which Nankints had stood became the mining camp La Esperanza.

But the Shuar did not take this lying down. The people displaced from Nankints took refuge in other Shuar communities nearby. They regrouped. And in November 2016, they fought back.

A group of Shuar men marched on La Esperanza, armed with spears, explosives, and rifles. The ExplorCobres employees and the police who were guarding the camp were taken by surprise and fled. But the Shuar's occupation of the camp was short lived. The next morning the police and military returned and drove them out.

In December, the Shuar counterattacked. This time, however, the police and military were ready for them. In the ensuing gunfight, José Luis Mejía, a policeman, was shot dead. To this day, no-one has been prosecuted for the killing and it remains unclear who is responsible. Diego Fuentes, interior minister at the time, said the bullet was fired from the forest using a specialist sniper rifle, but the Shuar insist they have no access to this kind of firepower (Watts, 2017).

Mejía's death enraged the then president Rafael Correa, who accused indigenous leaders of being in league with criminal and paramilitary groups (ibid.). He declared a state of emergency, essentially a two-month military occupation of the province of Morona Santiago, involving the use of armoured cars, tanks, drones, and helicopters.

The police and military raided houses in several communities in the area, looking for those suspected of involvement in Mejía's death. They eventually arrested six people – though charges were later dropped – while dozens of others in Shuar and mestizo communities in the area were criminalized, with offences ranging from inciting violence and possession of weapons to cattle theft (Ponce Ycaza, 2019).

Correa, a Kichwa-speaking mestizo, had come to power in 2007 as part of Latin America's so-called 'Pink Tide' (see Chapter 2) – and had enjoyed widespread backing from indigenous and social movements. Alongside Evo Morales, Correa was perhaps the leader most closely associated with the left-wing populism and drive towards regional integration led by Venezuela's Hugo Chávez. He was also the first Ecuadorian president to be sworn in according to native traditions, with leaders from Ecuador's indigenous movement blessing his 'Citizens' Revolution' (Caselli, 2011).

A new Constitution came into force in 2008, which defined Ecuador as a plurinational state, recognizing indigenous nationalities and Afro-Ecuadorians. Significantly, the Constitution also recognized nature as a legal subject with rights (see Conclusion). Though rights of nature laws exist in several countries worldwide, to this day, no other country has gone as far as Ecuador. This ground-breaking move was based on *sumak kawsay*, a principle of the Quechua peoples of the Andes which denotes a way of living in harmony with other human beings and with nature, in contrast to the capitalist logic of commodification and profit.

But it soon became apparent there was major discrepancy between what the new Constitution said on paper and what was to happen on the ground (Gordon and Webber, 2016, p. 214). In government, Correa pursued a programme of national-populist developmentalism, in which poverty relief and other social policies were funded by the aggressive expansion of extractive industries, particularly mining. This was accompanied by some highly combative discourse, with Correa frequently attacking his opponents on the left and within indigenous and environmental groups as 'childish' and occasionally even as 'extremists' or 'terrorists'.

By late 2009, the government was already in open conflict with Ecuador's indigenous movement, which, for demanding that Correa respect the principles of a Constitution his government had brought in only a year previously, was targeted with state repression, including serious criminal charges such as terrorism and sabotage. The state of emergency in Morona Santiago was one of the last – and most brutal – battles in this war between Correa and the very same indigenous groups who had helped bring him to power in 2007.

'It became a national conflict,' says Andrés Tapia. 'The leaders in the region were hunted down, they were among the country's most wanted people. There were rewards on offer for information leading to their arrest. They had charges hanging over them for years.'

San Carlos-Panantza was put on hold in late 2018 and the simmering conflict with the Shuar has meant that ExplorCobres has been unable to advance any further. In September 2022, the Constitutional Court ruled that the Shuar's right to consultation had been violated when the environmental licenses were granted, and ordered the government to carry out a consultation within six months. Given the degree of opposition the project has generated, this latest ruling makes it even less likely that it will go ahead (*El Universo*, 2022).

'According to the information we have at the moment, they haven't been able to move forward with it, despite all their efforts over recent years,' says Tapia. 'The reason for that is the resistance.'

***

While the global mining industry may not find life easy in Ecuador, its interest in the country is unlikely to diminish. Copper prices hit a record high in May 2021 and demand is set to increase further in the coming years. And with Ecuador having elected Guillermo Lasso – a right-wing banker – as president in 2021, companies may also feel that the current political climate is propitious in comparison to those of Peru and Chile, two of its main competitors.

Since Ollanta Humala stepped down in 2016, no Peruvian president has served a full term; there have been six presidents in less than seven years. In December 2022, leftist former schoolteacher Pedro Castillo was ousted, having attempted to dissolve Congress in order to prevent a vote on his impeachment. Though Castillo had pledged to reform Peru's mining industry, hoping to ensure more equitable distribution of the country's immense mineral wealth, he spent most of his brief presidency fighting for his survival and had none of the political capital required to take on such a powerful sector.

In Chile, recent social unrest and political upheaval has also brought uncertainty for the country's mining sector, with the vote to scrap the dictatorship-era Constitution (see Chapter 6) potentially leading to a tougher regulatory environment for the industry. This was further compounded in November 2021 with the election of the left-wing former student leader Gabriel Boric as president. Boric has signalled support for increased taxes and royalties on mining operations and promised that the most destructive projects would not be permitted to advance.

'To destroy the world is to destroy ourselves. We want no more sacrifice zones, no more projects which destroy our country, which destroy communities,' he said, in his first speech following his election victory (Reuters, 2021).

Boric's convictions are sure to be put to the test, as copper is not the only transition metal with which Chile is blessed (or cursed?). It is also the world's second biggest producer of a soft, ultra-light metal often referred to as the new 'white gold': lithium.

II. The Salar de Atacama, Chile: draining the desert

It takes about an hour to drive from the tourist hub of San Pedro de Atacama to the community of Peine, out in the desert in the Antofagasta region of northern Chile.

The scenery is unworldly. NASA uses the Atacama Desert as a testing ground for robots designed to conduct research on Mars and it's not hard to see why. It is the driest place on Earth outside the polar regions; parts of the desert are completely barren, devoid of vegetation or any other signs of life. Northern Chile is also home to 70 per cent of the global infrastructure for ground-based astronomy. The sky is almost always cloudless and there is very little light pollution, making conditions near perfect for stargazers (Johanson, 2021).

And yet, what is striking about the terrain around Peine is precisely that there *is* life, despite the arid conditions. Much of the landscape is covered with olive-green grasses and shrubs, irrigated by lagoons and underground water sources. On the horizon, mountains and volcanoes form an irregular grey-blue stripe; below them, an off-white band contrasts with the darker terrain of the desert. This is the Salar de Atacama, the largest salt flat in Chile and the third largest in the world after the Salar de Uyuni in Bolivia and Salinas Grandes in Argentina.

Peine, a small, dusty village home to around 300 people, is perched up on a hillside by the *salar*. A LAB journalist visited in 2018 to meet a group from the Consejo de Pueblos Atacameños (CPA), an umbrella organization representing 18 indigenous Atacameño communities in the surrounding region, including Peine. Its mission is to preserve their ancestral territory and culture, promoting their sustainable development via workshops and training, and conducting environmental monitoring and research.

Six people were waiting at the community centre, three men and three women. Two Wiphalas – the chequered rainbow banner representing the indigenous people of the Andes – stood planted in pots on the table.

It was Sergio Cubillos who did most of the talking. He was dressed casually: jeans, a T-shirt, a light beard of a few days darkening his face. Though by far the youngest member of the group, at the time of interview he was CPA president. After stepping down in early 2021, he ran for mayor of San Pedro de Atacama and then for deputy of the Antofagasta region, though in both cases without success.

'Our aim is to preserve our territory and subsist as a community,' he says. 'We want sustainable development, with access to quality education and a healthcare system that meets our needs. We're not against the country continuing to progress and develop, but we think there needs to be a change in the vision of development.'

The territory of the Atacameños has long been under pressure from the mining industry, particularly the copper mines Escondida – operated by the Anglo-Australian miner BHP – and Zaldívar, a joint venture between the Chilean company Antofagasta Minerals[2] and Canada's Barrick Gold. But in the last two decades, the Atacameños have watched in dismay as the *salar* has become one of the main global hubs for extraction of yet another commodity: lithium. Beneath the Salar de Atacama lies an estimated one third of the entire global supply.

Lithium is an essential component of lithium-ion batteries, which store large quantities of energy relative to their size and weight. They will be a key means of energy storage as the world transitions away from fossil fuels; demand is expected to skyrocket in the next few years. However, with the damage that lithium mining is doing to the *salar* becoming increasingly obvious, the Atacameños are determined to put the brakes on any further expansion.

The true cost of e-mobility

Antofagasta is the very heart of the Chilean mining industry, providing many of the raw materials which have driven Chile's great export booms.

In the 19th and early 20th centuries there was saltpetre – the original 'white gold' – which was widely used as a fertilizer and in gunpowder. Then came copper, which has long been Chile's main export. Antofagasta is home to some of the biggest mines in the world, such as Chuquicamata, the 20th century's most prolific copper mine (Brown, 2012, p. 132), as well as Escondida, currently the world's largest copper-producing operation. Aside from these immense copper projects, there are also gold, iron, and nitrate mines.

Today, Chile's copper industry alone accounts for more than half of Chilean exports (Santander Trade Markets, 2022) and around ten per cent of national GDP (Barría, 2021). But while the Chilean state has always treated Antofagasta as a cash cow, for the Atacameños it is their ancestral home. Traditionally, they were subsistence farmers, herding alpacas, and llamas, then, after the arrival of the Spanish, sheep, goats, and mules. They also grow crops, such as quinoa, maize, squash, and chillies (Museo Chileno de Arte Precolombino, n.d.).

Their coexistence with mining has always been uneasy and as the industry has expanded, and its impacts have become harder to ignore, Atacameño resistance has hardened. This has been informed by a growing sense of their identity as indigenous communities and a greater awareness of their rights.

'The first time Escondida came, offering us a pittance, we didn't know what an indigenous community was, because we were all just Chileans, so we made do with what they gave us,' says Sara Plaza, one of the women at the meeting in Peine. 'Later we began to understand what an indigenous community is, and we began to fight the mining company.'

They point to the destruction of the Salar de Punta Negra, further to the south, as a cautionary tale. For years, BHP had been pumping water from beneath the salt flat for use at Escondida. Though it permanently ceased this practice in 2017, the Atacameños argue that it was too late. Both Escondida and the Salar de Punta Negra are over 100 km from Peine, which, Cubillos says, is part of the reason BHP was able to get away with it for so long. But Peine is located on the edge of the Salar de Atacama and the impacts of mining there are much more immediate.

'Water was extracted from the Salar de Punta Negra for more than 20 years … They're no longer taking water, but the Salar has been destroyed,'

Photo 7.2 Sara Plaza on the plains below Peine / © Grace Livingstone 2019

says Cubillos. 'And this is exactly what we don't want to happen to the Salar de Atacama.'

But it is in grave danger – and with the need to decarbonize the global economy becoming ever more urgent, the future of the *salar* is far from certain.

Lithium-ion batteries are found in everything from laptops and mobile phones to electric scooters and e-cigarettes. But these technologies, though widespread, use only small quantities of the metal. What really threatens to send demand through the roof is the expanding market for electric vehicles (EVs), which are far more lithium intensive. The battery of a Tesla Model S contains about twelve kilos, for instance (Katwala, 2018).

Road vehicles in the UK account for about 19 per cent of national CO_2 emissions, a figure which must be reduced if the target of reaching net zero by 2050 is to be met. Given that sales of new petrol and diesel vehicles are to be banned from 2030, mass uptake of EVs is a central part of the plan. Nor is the UK alone. The European Union has proposed a ban on new fossil fuel-powered vehicles from 2035 and various U.S. states are introducing similar legislation, including California and New York.

The issue is not only that EVs are metal-intensive. Our national energy networks, in their current form, are not built for widespread EV use. Existing power networks will have to be modernized, to build in the capacity to store and regulate the vast amounts of – hopefully renewable – energy required. This means more metals.

For the mining industry, this represents a major business opportunity. But for the communities who live near the Andean *salares,* and the scientists who study these unique ecosystems, the spiralling global demand for lithium is serious cause for concern.

Cristina Dorador is a biologist at the Universidad de Antofagasta and was one of the delegates elected to the constitutional assembly, the body which was responsible for drafting Chile's new constitution.

'In the Atacama Desert a range of different aquatic environments can be found, the remnants of lakes from millions of years ago,' she wrote in an article for LatinAmericanScience.org. 'Today, they're home to enormous biodiversity, including endemic species of plants and animals, as well as microorganisms which are a living testimony to how life on this planet is possible even in extreme conditions' (Dorador, 2014).

'What will become of future generations?'

This arid, altiplano region, spreading south-east into Argentina and north-east into Bolivia, is sometimes called the 'Lithium Triangle'; it is thought to hold around 75 per cent of global supply. Chile is by far the leading producer of the three, though Argentina is trying to catch up. Bolivia is thought to hold the biggest reserves of all, though for various political, economic, and technical reasons its nascent industry has yet to take off.

In all three countries, the lithium is contained not within rock, but brine: mineral-rich groundwater beneath salt flats. Companies sink wells, then pump the brine into vast pools on the surface and let the sun evaporate the water. What's left – usually a combination of lithium, potassium, manganese, and other minerals – is filtered and left to evaporate once more. The whole process takes between 12 and 18 months, after which time the lithium carbonate can be extracted (Ahmad, 2020).

The companies have long claimed that the evaporation process is clean and has no impact on the wider ecosystem: the brine is fit neither for human consumption, nor for agriculture, so why does it matter if this water is lost to the atmosphere?

But things may not be so simple. The debate centres on the relationship between the brine underneath the salt flat and the freshwater systems at its perimeter.

Mariana Cervetto is a hydrogeologist who has worked with the Atacameños. 'In the hydrological models that the mines present, they say that pumping brine will never, ever have an effect on the normal function of the other aquifer,' she says. 'That no matter how much brine they pump, they won't affect these lakes. And in my objective, technical opinion, I wouldn't be so sure.'

Thea Riofrancos, a political scientist based at Providence College and a specialist on extractive industries in Latin America, likens it to an arm-wrestling match (Riofrancos, 2019). On one side is the brine, which is much denser than freshwater due to its high mineral content. On the other is the freshwater, which comes from snowmelt up in the Andes and which gravity drags down towards the centre of the *salar*.

Under normal circumstances, these two opposing forces hold each other in check. But when the brine is removed, there is nothing to stop the freshwater from flowing down into the salt flat – draining the surrounding lagoons and wetlands and leaving the indigenous communities around the edge of the *salar* with insufficient water for domestic and agricultural use.

A growing body of scientific evidence suggests this is happening, backed up by anecdotal evidence from the communities.

'We have some grasslands with a well, and it really hurts me, because it used to be a big grassy plain with a well full of water and now it's dry, it's dry!' says Sara Plaza, with tears in her eyes. 'I went there recently and it hurts, because we've been fighting really hard, but now this grassland has all dried up, there's very little water.'

'It's terrible for us elders because we've been fighting for so long, but now our wetland is dry and what will happen afterwards? Will there be a future? What will become of future generations?'

Ultimately, how much brine lies beneath the salt flat remains unknown, as does the precise nature of the relationship between the two water systems. Studies still haven't been done in enough detail and over enough time to

Photo 7.3 Area of remaining wetland on plains below Peine / © Grace Livingstone 2019

draw any firm conclusions about the long-term effects of lithium mining on the *salar*.

'It's very difficult to know because here in Chile we don't have basic information,' says Cervetto. 'The information is produced by the same companies when they have to present their environmental impact studies. It's the same company who presents the information, the same company who decides what kind of monitoring they'll do and who defines the loss.'

In other words, it's yet another example of how, all too often, mining companies are allowed to regulate themselves, even when operating in areas of great ecological sensitivity.

'Our position is that no companies should be allowed to set up within our territory while the necessary studies still haven't been done on the health of the *salar*,' says Sergio Cubillos. 'But we know that this could go on for much longer. People say the lithium boom could go on for forty or fifty years. We will be keeping a close eye on the situation.'

Legal battles

Since 1980, just two companies have been extracting lithium from the Salar de Atacama: Albemarle, an American chemical company, and the Sociedad Química y Minera de Chile (SQM), a former state-owned company that was privatized during the military dictatorship (1973–1990) and handed to Julio Ponce Lerou, son-in-law of General Pinochet.

SQM is controversial not only for its links with the dictatorship. It is at the very centre of one of the biggest corruption scandals in recent Chilean history, having for years illegally channelled campaign funding to politicians from across Chile's political establishment. Ponce Lerou stepped down as company chairman in 2015 amidst these allegations, but retains a 30 per cent share in the company, making him the largest single shareholder.

The Consejo de Pueblos Atacameños has an agreement with Albermarle, under which the communities receive 3.5 per cent of the profits from lithium sales and both parties participate in a programme of environmental monitoring. The CPA is able to hire specialists to monitor the impacts of extraction – such as Mariana Cervetto – as well as lawyers to represent the communities before the courts and professionals in other areas.

However, no such agreement exists with SQM, which in 2016 found itself in trouble with Chile's environmental authorities over several infractions on the Salar de Atacama, including de-activating wells that were supposed to be used for monitoring purposes and extracting more brine than was permitted (Carrere, 2019). To avoid losing its licenses, SQM submitted a $25 million mitigation plan, which was accepted by the authorities.

But for the Atacameños, the plan was a distraction which did little to repair the damage to the *salar*. They went to court, requesting that the plan be scrapped – and they won. As of August 2021, SQM had been told to submit a new plan, or risk seeing its hopes for further expansion scotched. For the

Atacameños, this was an important victory – but they would like to see SQM removed from the Salar de Atacama altogether.

'We will make every effort to see SQM's [environmental permits] revoked,' says Cubillos. 'The damages committed by this company are immeasurable and they must assume their responsibility.' (Sherwood, 2020)

The CPA also went to court over an agreement between SQM and the government, which not only gave the company the right to continue extracting lithium until 2030, but allowed it to triple its permitted production quotas. The CPA argue that they weren't consulted, in violation of Convention 169 of the International Labour Organization (ILO). Though the Supreme Court threw the case out in 2019, the CPA filed a complaint with the Inter-American Court of Human Rights (IACHR), which is still pending. In theory, rulings by the IACHR are binding on national states, though the level of non-compliance is high.

Other companies have so far failed to extract anything from the Salar de Atacama. They have complained about a lack of clarity regarding the financial and environmental conditions required by regulators, while the opposition from the Atacameños has also been a factor. Even so, in the long term, Cubillos is not optimistic.

'The companies have a big interest in setting up here, though this has been dampened by the actions not just of the Consejo, but the communities too,' he says. 'But unfortunately, in Chile, the legislation is designed so that these kinds of projects are approved swiftly, with barely any consultation or any possibility of having them rejected outright.'

Though these legal challenges are time consuming and don't always bring the desired results, they remain the *modus operandi* of the Atacameños, alongside their educational and environmental work. They have avoided direct action like roadblocks or attacks on company property, but with so much at stake, there is still a latent threat of violence.

'Here you don't see violence, or any police, but I think if we were to resist the way the Mapuches have, we would have the same level of repression, or even worse, because this is where the country's economic power and interest is located,' says Cubillos, referring to the struggle of Chile's largest indigenous group against the advance of forestry and agriculture in the south of the country.

'Here you have CODELCO, Escondida, lithium...all of Chile's economic interests are here in this region. We would need only to hold one hunger strike, or close down San Pedro de Atacama, and they would come ...'

***

While Chile's share of the market has fallen in recent years, it remains an extremely attractive destination for companies looking to get into lithium mining. The brine beneath the Salar de Atacama is higher grade than that found in Bolivia and Argentina, and the environmental conditions for evaporation are ideal: constant sunshine during the day, virtually no rain, and hot, dry winds (Sherwood, 2019). Chile's location on the Pacific coast of South America is also convenient: the lithium that comes from the Salar de Atacama is taken directly to the port city of Antofagasta and shipped off to global markets. With demand projected to skyrocket in the next few years, there are potentially vast fortunes to be made from continuing to drain the *salar*.

Much may depend on the new constitution, and whether it substantially overhauls Chile's completely privatized system of water management (see Chapter 6). Currently, the brine beneath the *salar* is treated as a mineral, not as freshwater. But with growing evidence of the interaction between the brine and nearby freshwater, this could change.

Moreover, the brine is not the only water used in lithium mining. The companies still use freshwater to clean machinery and infrastructure and to wash and process the lithium carbonate. Producing a single tonne of lithium carbonate typically consumes over two million litres of water (UNCTAD, 2020).

'We are in a desert, the driest desert in the world,' says Cubillos. 'So why would we privilege the use of water by a mining company, when we're unable to meet the needs of the community and its development?'

III. The myth of sustainable mining

Mining companies have long employed greenwashing: the deliberate appropriation of the language and imagery of environmentalism and sustainability to persuade the public that – to borrow a phrase from Anglo American Chile – 'mining and the environment can coexist harmoniously' (Anglo American-Chile, 2019). These efforts have intensified considerably in recent years, coinciding with the increasing global alarm about climate change, deforestation, and biodiversity loss.

But while mining companies may masquerade as the gatekeepers to a sustainable, low-carbon future, historically they have been a major cause of the problem. Mining operations require a lot of energy, which is often produced by burning fossil fuels. They deforest large areas. Vehicles and machinery emit greenhouse gases. And if indirect emissions are considered – basically the burning of coal by companies the industry supplies – the contribution of the mining industry to global carbon emissions could be as high as 28 per cent (Delevingne, et al., 2020).

Mining companies also continue to extract other metals and minerals that have no green application, and even, in the case of gold, limited industrial use. Most gold is used to make jewellery or is converted into ingots and stored in bank vaults as a low-risk investment and protection against economic crisis.

Much of the language of greenwashing is borrowed from NGOs and international institutions such as the UN and World Bank – particularly that associated with the UN's Sustainable Development Goals (SDGs). These institutions have also participated in the greenwash, promoting mining as a source of development for low-income countries and the key to a sustainable future, often with little nuance.

The Future We Want – the declaration of the UN Conference on Sustainable Development, held in Rio de Janeiro in 2012 – argues that 'mining offers the opportunity to catalyze broad-based economic development, reduce poverty and assist countries in meeting internationally agreed development goals ... when managed effectively and properly.' It also states that 'we recognize that governments need strong capacities to develop, manage, and regulate their mining industries in the interest of sustainable development.' (United Nations, 2012, p. 59)

The assumption seems to be that mining can be made sustainable, or at least, that it can be reconciled with the demands of sustainable development. But what does this really mean? Is 'sustainable mining' not an oxymoron?

The answer is yes, if we define mining as the extraction of non-renewable metals and minerals from the ground, for then mining is indeed an inherently unsustainable activity. Even if we interpret the concept less literally, there is little consensus on what exactly 'sustainable mining' might mean in practice. Broadly speaking, industry discourse on sustainability, alongside that of the UN and other global institutions, suggests it is about improving Environmental, Social and Governance (ESG) standards: from health and safety to gender equality; from reducing water consumption to education and training.

But it is telling that many of the initiatives often included within the category of 'sustainability' – healthcare, education, youth development and so on – actually have little or nothing at all to do with mining as such. In other words, the fact that mining companies feel the need to offer these 'sustainable' extras which, in theory, will endure beyond the active life of the mine, is a tacit admission that the activity itself is *not* sustainable (Godfrid, 2018, p. 213). As the Mexican government put it, in a guide issued to support the industry, 'Mining companies have become more conscious of the need to promote community development, as a way of making their business sustainable.' (Secretaría de Economía, n.d., p. 15)

Much emphasis is placed on community relations: obtaining consent from local people for operations; involving them in decision making; and leaving a positive legacy after mine closure. But many communities in Latin America see large-scale industrial mining as incompatible with their traditions, livelihoods, and ways of life; they are not interested in negotiating. In such situations, the only way for companies to obtain a foothold is through subterfuge and violence.

'There are very few communities that want mining. Those that do tend to have been persuaded by the propaganda of mining companies and governments,' says César Padilla, of the Latin American Observatory of Mining Conflicts (OCMAL). 'When they realize what mining entails, they change their mind. But by that point it's too late and there's no turning back.'

A charitable assessment would be that 'sustainable mining' encapsulates the industry's reckoning with itself, a recognition of past failures and a commitment to do better in the future. A more critical view would be that it is simply propaganda, the appropriation of a term with green connotations in order to legitimize an activity that continues to cause enormous harm.

The 'copper crunch'

The unsustainability of mining in its current form – even in the medium term – becomes clearer still if we take a closer look at one key metal: copper.

Copper is sometimes overlooked in discussions of green technology, given its widespread conventional uses, but there will be no energy transition without it. It is needed for wiring and piping in new infrastructure; it is used in solar panels and wind turbines; it is a key component of 5G technology; and it is essential for EVs, which contain about four times as much copper as a fossil fuel-powered vehicle (Mills, 2020).

In an analysis of the UK government's proposals on the EV rollout, a team of scientists at the Natural History Museum crunched the numbers. Their conclusion was that if we were to replace all the UK's 31.5 million cars and vans with electric vehicles, it would consume the equivalent of 12 per cent of the world's entire copper output – not to mention three-quarters of global lithium production – for 2018 (Natural History Museum, 2019).

Of course, just because new petrol and diesel vehicles will no longer be sold in the UK from 2030, it doesn't mean that Brits are going to stop driving them overnight. Still, when we consider that the EU, the United States, China, Japan, and other countries will be attempting to phase out fossil fuel vehicles at the same time, it's clear that the pressure on global copper supply will be immense.

According to a conservative estimate by mining analyst Rick Mills, to achieve 30 per cent global EV coverage, on top of the projected consumption for all other uses of copper, global production needs to more than double in the next 20 years. Where is all this copper going to come from? Many of the world's major copper mines are running out of ore. Over 200 copper mines are expected to come offline by 2035 and there are not enough new mines in the pipeline to pick up the slack (Mills, 2020).

In traditional copper mining countries such as Chile and Peru, mining is now a highly sensitive issue, particularly when it comes to water. People are increasingly weighing up the costs and benefits and deciding that the sums don't add up, hence the calls for new taxes and royalties. Big new copper projects are likely to encounter stiff opposition at the local level, meaning they may not be possible at all without riding roughshod over people's human rights. There is still untapped copper underground, in countries such as the United States, Russia, Afghanistan and the Democratic Republic of Congo – but this implies expanding the industry into virgin ecosystems with no history of mining, or, in some cases, into areas ravaged by years of armed conflict.

Finally, as prices rise, there is also the danger that increasingly low-grade deposits of copper become viable to mine, with all the negative environmental impact that implies. In the words of the mining entrepreneur Robert Friedland:

> ... it's just not that good for Mother Earth to be grinding up enormous volumes of 0.3 of 1 per cent copper. So the first question is, well where are you getting your electricity? How much water are you using? Where are you putting these oceans of tailings, and are these oceans of tailings safe for all eternity when you put them behind a dam? What's the life of that dam? Twenty years? Fifty years? A hundred years? Ten thousand years? Will the tailings dams last as long as the Pyramids, for example? (Saywell, 2021)

***

This is not to say that mining is uniquely evil or should never happen. There are many industries which are also unsustainable in their current form and at their current pace: the energy sector, aviation, and cattle farming, to name just three.

Of course, we need metals and minerals. Without them we have no electricity, no central heating or hot water, no motorized transport, and no communications technologies. And yes, combating global warming and providing heat and light for a growing global population will not be possible without metals and minerals too.

'Our society needs minerals,' says the Brazilian environmental engineer Bruno Milanez, 'but if we're to move mineral extraction towards something that can be called "sustainable development", then we need to rethink the scale, the methods, and the pace of extraction, as well as the use and the waste of these materials' (Fachin, 2015, p. 35).

For Joan Kuyek, a founding member of the NGO MiningWatch Canada, 'Corporations are externalizing machines; they seek to offload all the costs they can from their corporate account to the environment, individuals, and the public sector' (Kuyek, 2019, p. 139). The current system, based largely on the principle of industry self-regulation, makes it easy for companies to do just this. Any serious attempt to reconcile mining with sustainable development would therefore have to bring an end to self-regulation and force companies to account for the true social and environmental costs of their operations.

We also need to drastically expand our capacities for re-use and recycling. Metals do not lose their mechanical or metallurgical properties when recycled and in theory, provided they can be safely extracted from the products in which they have been used, could be recycled almost endlessly (ibid., p. 142). In the case of the strategic or green minerals, it is estimated that expanded recycling could reduce demand by between 25 and 55 per cent. The problem is that current policies incentivize new extraction rather than re-use or recycling (Sampat, 2021).

The overall aim should be what the Uruguayan researcher Eduardo Gudynas calls 'indispensable extractivism'. This entails a massive downsizing of all extractive industries (including mining), so that 'the only ones left are those that are genuinely necessary, meet social and environmental conditions, and are directly linked to national and regional economic chains' (Gudynas, 2011, p, 175).

So, to return to the question posed at the beginning of this chapter: can we dig ourselves out of this hole?

The kind of exponential increase in mining production as projected by the World Bank will involve vastly expanding the mining frontier in Latin America and elsewhere, devastating hitherto pristine areas and endangering the conditions communities depend on for their survival. There is a strong element of climate *injustice* at work here: these communities are at risk of being sacrificed for an energy transition in which they will participate little, to solve a problem to which their contribution is negligible.

That said, realistically, there will have to be some digging, particularly in the short term. However, tackling climate change will involve far more than just swapping petrol cars for EVs, or even fossil fuels for renewable energy.

A 2018 report by the Intergovernmental Panel on Climate Change states that limiting global warming to 1.5 °C will involve 'rapid, far-reaching and unprecedented changes in all aspects of society' (IPCC, 2018). It is about how we live and work, how we organize our cities, where we go on holiday, the food we eat, the clothes we wear. It will involve a radical reassessment of how we conceive of our quality of life, a shift away from a model of private ownership and accumulation towards one based on public ownership and sharing, with the aim of making the best possible use of resources in order to satisfy people's real needs.

We need to be honest about the scale of the transformation involved. This is why the whole discourse around 'sustainable mining' is so unhelpful: by implying that we can simply replace massive fossil fuel extraction with mining, no strings attached, it precludes thinking about the radical, holistic changes that we so urgently need to make.

We must find ways to make do with less. This isn't a call for a return to some pre-industrial age; on the contrary, this is our chance to rise to the challenges of the 21st century, and to free ourselves from a wasteful, inefficient, and unjust model of consumption which is rapidly destroying the conditions necessary for human life on this planet. A sustainable future will be fairer and more rational, or it will not be sustainable at all.

Notes

1. Ecuador's currency has been the U.S. dollar since the year 2000. While this has helped to maintain a degree of economic stability, Ecuador cannot set its own monetary policy, which leaves the country vulnerable to fluctuations in the competitiveness of its exports and the prices of imports as the dollar rises and falls.

2. Antofagasta is sometimes described as a British company, being registered in England and Wales and having an office in London. However, while it may raise its capital in the UK, the company and all of its mining operations are based in Chile, and it is controlled by the Luksic family, one of Chile's wealthiest families.

References

All references to web-based material were checked and still available in November 2022 unless otherwise stated.

All references are listed, with clickable links for your convenience, on the page for this chapter on the Heart of Our Earth website: <https://lab.org.uk/the-heart-of-our-earth/>

Ahmad, S. (2020) 'The Lithium Triangle: Where Chile, Argentina, and Bolivia Meet'. [online] *Harvard International Review*. Available at: <https://hir.harvard.edu/lithium-triangle/>.

Alvorado, A. C. (2022) 'Ecuador: proyecto minero Mirador genera nuevas amenazas de desalojo en Tundayme'. [online] Mongabay. Available at: <https://es.mongabay.com/2022/04/ecuador-proyecto-minero-mirador-genera-amenazas-de-desalojo/>.

Anglo American (2021) 'Climate Change'. [online] Available at: <https://www.angloamerican.com/sustainability/environment/climate-change>.

Anglo American-Chile (2019) *Anglo American-Chile on Twitter: "Minería y medio ambiente pueden coexistir de manera armónica."*. [online] Available at: <https://twitter.com/AngloAmericanCL/status/1113498172305883136?s=20>.

Barría, C. (2021) 'Cómo el enorme aumento del precio del cobre encendió el debate en Chile sobre el aporte de las mineras privadas a las arcas públicas'. [online] BBC News Mundo. Available at: <https://www.bbc.com/mundo/noticias-56247275>.

Bermúdez Liévano, A. (2019) 'Indigenous communities take legal action over Ecuador's largest mine'. [online] Diálogo Chino. Available at: <https://dialogochino.net/en/extractive-industries/28120-indigenous-communities-take-legal-action-over-ecuadors-largest-mine/>.

Brown, K. (2012) *A History of Mining in Latin America. From the Colonial Era to the Present*. Albuquerque: University of New Mexico Press.

Carrere, M. (2019) 'Chile: detienen proceso sancionatorio de SQM acusada de graves infracciones ambientales'. [online] Mongabay. Available at: <https://es.mongabay.com/2019/01/chile-detienen-sanciones-por-danos-ambientales-en-salar-de-atacama/>.

Caselli, I. (2011) 'Ecuador President Rafael Correa loses indigenous allies'. [online] BBC News. Available at: <https://www.bbc.co.uk/news/world-latin-america-12944231>.

Colectivo de Investigación y Acción Psicosocial (2017) *La herida abierta del Cóndor. Vulneración de derechos, impactos socioecológicos y afectaciones psicosociales provocados por la empresa minera china EcuaCorriente S.A. y el Estado ecuatoriano en el Proyecto Mirador*. [online] Quito. Available at: <https://investigacionpsicosocial.files.wordpress.com/2017/02/herida-abierta-del-cc3b3ndor.pdf>.

Corriente Resources Inc. (n.d.) 'Corriente Resources Inc. | Corporate | Overview'. [online] Available at: <http://www.corriente.com/corporate/corporate_overview.php>.

Delevingne, L., Glazener, W., Grégoir, L. and Henderson, K. (2020) 'Climate risk and decarbonization: What every mining CEO needs to know'. [online] McKinsey Sustainability. Available at: <https://www.mckinsey.com/business-functions/sustainability/our-insights/climate-risk-and-decarbonization-what-every-mining-ceo-needs-to-know>.

Dorador, C. (2014) 'El universo microbiano de Atacama: la riqueza de lo que no se ve'. [online] Latin American Science. Available at: <http://latinamericanscience.org/spanish/2014/07/el-universo-microbiano-de-atacama-la-riqueza-de-lo-que-no-se-ve/>.

El Universo (2022) 'Corte Constitucional sienta precedente para que el Estado cumpla con consulta previa a pueblos indígenas antes de implementar proyectos extractivos'. [online] Available at: <https://www.eluniverso.com/noticias/politica/corte-constitucional-sienta-precedente-para-que-el-estado-cumpla-con-consulta-previa-a-pueblos-indigenas-antes-de-implementar-proyectos-extractivos-nota/>.

Fachin, P. (2015) 'Exportação de minério e a opção brasileira pela crise permanente. Entrevista especial com Bruno Milanez'. *Cadernos IHU em formação. Mineração e o impulso à desigualdade: impactos ambientais e sociais*, [online] (48), pp. 33–37. Available at: <http://www.ihu.unisinos.br/images/stories/cadernos/formacao/48ihuemformacao.pdf>.

Fundación Pachamama (2020) *La narrativa de que el modelo minero sacará al Ecuador de la crisis desatada por la pandemia del Coronavirus ...* [video] Available at: <https://www.facebook.com/watch/?v=247042456734973>.

Godfrid, J. (2018) 'La implementación de iniciativas de responsabilidad social empresaria en el sector minero. Un estudio a partir de los casos Alumbrera y Veladero'. In: L. Álvarez Huwiler and J. Godfrid, eds., *Megaminería en América Latina. Estados, empresas transnacionales y conflictos socioambientales*. [online] Buenos Aires and Quilmes: Centro Cultural de la Cooperación Floreal Gorini and the Universidad Nacional de Quilmes, pp. 199–228. Available at: <https://www.researchgate.net/publication/334635299_La_implementacion_de_iniciativas_de_responsabilidad_social_empresaria_en_el_sector_minero_Un_estudio_a_partir_de_los_casos_Alumbrera_y_Veladero/link/5d3719aa299bf1995b439c1f/download>.

Gordon, T. and Webber, J. (2016) *Blood of Extraction. Canadian imperialism in Latin America*. Halifax, Nova Scotia and Winnipeg, Manitoba: Fernwood Publishing.

Gudynas, E. (2011) *Transitions to post-extractivism: directions, options, areas of action*. [online] Transnational Institute. Available at: <https://www.tni.org/files/download/beyonddevelopment_transitions.pdf>.

Hui, N. (2019) 'Few options left for local communities opposing Ecuador's largest copper mine'. [online] China Dialogue. Available at: <https://china-dialogue.net/en/business/11361-few-options-left-for-local-communities-opposing-ecuador-s-largest-copper-mine/>.

IPCC (2018) 'Summary for Policymakers of IPCC Special Report on Global Warming of 1.5°C approved by governments'. [online] Available at: <https://www.ipcc.ch/2018/10/08/summary-for-policymakers-of-ipcc-special-report-on-global-warming-of-1-5c-approved-by-governments/>.

Johanson, M. (2021) 'What it's like to visit Mars on Earth'. [online] CNN. Available at: <https://edition.cnn.com/travel/article/atacama-desert-chile-mars/index.html>.

Katwala, A. (2018) 'The spiralling environmental cost of our lithium battery addiction'. [online] *WIRED UK*. Available at: <https://www.wired.co.uk/article/lithium-batteries-environment-impact>.

Kuyek, J. (2019) *Unearthing Justice. How to protect your community from the mining industry*. Toronto: Between The Lines.

McGrath, M. (2020) 'Climate change: Temperature analysis shows UN goals 'within reach''. [online] BBC News. Available at: <https://www.bbc.co.uk/news/science-environment-55073169>.

Mills, R. (2020) 'Copper, the most critical metal'. [online] Mining.com. Available at: <https://www.mining.com/web/copper-the-most-critical-metal/>.

Morán, S. (2019) 'Tundayme sin derecho a ser indígena'. [online] Plan V. Available at: <https://www.planv.com.ec/historias/sociedad/tundayme-sin-derecho-ser-indigena>.

Museo Chileno de Arte Precolombino (n.d.) 'Chile's indigenous peoples | Atacameño | Economy'. [online] Available at: <http://precolombino.cl/en/culturas-americanas/pueblos-originarios-de-chile/atacameno/#/economia/>.

Natural History Museum (2019) 'Leading scientists set out resource challenge of meeting net zero emissions in the UK by 2050'. [online] Available at: <https://www.nhm.ac.uk/press-office/press-releases/leading-scientists-set-out-resource-challenge-of-meeting-net-zer.html>.

Pérez, A. (2019) 'Ecuador: tres proyectos mineros acechan la riqueza ambiental de la Cordillera del Cóndor'. [online] Mongabay. Available at: <https://es.mongabay.com/2019/05/cordillera-del-condor-en-ecuador-tres-proyectos-mineros-la-acechan/>.

Ponce Ycaza, I. (2019) 'The Ghost of Nankints'. [online] Diálogo Chino. Available at: <https://dialogochino.net/en/extractive-industries/26258-the-ghost-of-nankints/>.

Reuters (2021) 'Boric dice que se opondrá a proyecto minero de hierro y cobre Dominga'. [online] *La República*. Available at: <https://www.larepublica.co/globoeconomia/gabriel-boric-dice-que-se-opondra-a-proyecto-minero-de-hierro-y-cobre-dominga-3279644>.

Riofrancos, T. (2019) 'What Green Costs'. [online] *Logic Magazine*. Available at: <https://logicmag.io/nature/what-green-costs/>.

Rio Tinto (2021) 'Climate Change'. [online] Available at: <https://www.riotinto.com/sustainability/climate-change>.

Roberts, D. (2020) 'The sad truth about our boldest climate target'. [online] Vox. Available at: <https://www.vox.com/energy-and-environment/2020/1/3/21045263/climate-change-1-5-degrees-celsius-target-ipcc>.

Ruiz Leotaud, V. (2020) 'Corriente Resources' mining camp in Ecuador set on fire'. [online] Mining.com. Available at: <https://www.mining.com/corriente-resources-mining-camp-in-ecuador-set-on-fire/>.

Sampat, P. (2021) *Making clean energy clean, just & equitable*. [Webinar] MiningWatch Canada. Available at: <https://www.youtube.com/watch?v=-ue_kUN8-D0>.

Sánchez-Vázquez, L. and Leifsen, E. (2019) 'Resistencia antiminera en espacios formales de gobernanza: El caso de CASCOMI en Ecuador'. *European Review of Latin American and Caribbean Studies*, (108), 65–86.

Santander Trade Markets (2022) 'Chilean foreign trade in figures'. [online] Available at: <https://santandertrade.com/en/portal/analyse-markets/chile/foreign-trade-in-figures>.

Saywell, T. (2021) 'CRU World Copper Conference: Friedland on the new world order'. [online] Mines and Communities. Available at: <http://www.minesand communities.org/article.php?a=14568>.

Secretaría de Economía (n.d.) *Guía de Ocupación Superficial. Alianzas Estratégicas para la Promoción y el Desarrollo de la Competitividad del Sector Minero Mexicano*. [online] Available at: <https://www.economia.gob.mx/files/comunidad_negocios/industria_comercio/informacionSectorial/minero/guia_de_ocupacion_superficial_0414.pdf>.

Sherwood, D. (2019) 'Chile, once the world's lithium leader, loses ground to rivals'. [online] Reuters. Available at: <https://www.reuters.com/article/us-chile-lithium-analysis-idUSKCN1T00DM>.

Sherwood, D. (2020) 'Indigenous groups in Chile's Atacama push to shut down top lithium miner SQM'. [online] Reuters. Available at: <https://www.reuters.com/article/us-chile-lithium-sqm-idUSKCN25A2PB>.

UNCTAD (2020) 'Developing countries pay environmental cost of electric car batteries'. [online] Available at: <https://unctad.org/news/developing-countries-pay-environmental-cost-electric-car-batteries>.

United Nations (2012) *The Future We Want. Outcome document of the United Nations Conference on Sustainable Development*. [online] Available at: <https://sustain-abledevelopment.un.org/content/documents/733FutureWeWant.pdf>.

Warnaars, X. (2013) 'Territorial transformations in El Pangui, Ecuador: Understanding how mining conflict affects territorial dynamics, social mobilisation, and daily life'. In: A. Bebbington and J. Bury, eds., *Subterranean Struggles. New dynamics of mining, oil, and gas in Latin America*. Austin, TX: University of Texas Press, pp. 149–171.

Watts, J. (2017) 'Amazon land battle pits indigenous villagers against might of Ecuador state'. [online] *The Guardian*. Available at: <https://www.theguardian.com/world/2017/mar/19/ecuador-indigenous-shuar-el-tink-mining-land-dispute>.

World Bank (2020) 'Mineral Production to Soar as Demand for Clean Energy Increases'. [online] Available at: <https://www.worldbank.org/en/news/press-release/2020/05/11/mineral-production-to-soar-as-demand-for-clean-energy-increases>.

CHAPTER 8

Conclusion: An end to business as usual

It is hard to imagine what the world would look like today had it not been for mining in Latin America. It drove Spanish and Portuguese imperialism and helped to define the borders of their colonies. It sparked mass movements of people across the globe, and then from coastal areas of the Americas to the interior. It produced a glittering stream of wealth which flowed across the Atlantic, to Spain and Portugal, northern Europe, and beyond, monetizing the global economy and laying the foundations for the development of modern capitalism.

Yet for Latin America, the prosperity generated by mining has often been ephemeral. The benefits have accrued mostly to relatively small numbers of people associated with the industry, while the environmental and human costs have been appalling. Despite the region's immense mineral wealth, its economic performance has lagged behind most of the rest of the world over the last 200 years, while efforts to diversify economies and create more sustainable industries have met with only limited success.

Since the 1990s, in line with the recommendations of the Washington Consensus, Latin American governments have instead reinforced and expanded an economic model based on the extraction of primary commodities for export. Mining, of course, is only one part of a much bigger picture. During this period, hydrocarbon extraction, agribusiness, electricity generation (including renewables, such as hydroelectric plants and windfarms), major works of infrastructure, and tourist developments, have all brought about extraordinary and unprecedented changes in the geography of Latin America. The most alarming consequence of this is the ongoing and accelerating destruction of the Amazon, a ten-million-year-old forest of which, in little over half a century, nearly a fifth has been destroyed (Sandy, n.d.).

Inevitably, these environmental changes bring social and cultural impacts in their wake. Lifestyles which have changed little in centuries are disappearing in a generation or two. Displacement in rural areas has led to the unplanned and uncontrolled growth of towns and cities, with migrants often ending up in peripheral neighbourhoods and informal settlements like *favelas*, where they suffer from substandard housing; lack of essential infrastructure such as sewage, water, and electricity networks; poor access to public services and employment; and exposure to organized and violent crime.

Despite growing awareness of these impacts, the expansion of the mining frontier looks set to continue apace. In the last twenty years, the price of silver has increased more than fivefold, those of gold and copper more than sixfold.[1] While the Covid-19 pandemic depressed prices initially for many metals, they have rebounded strongly. The global energy transition – now an area of policy consensus across Europe, China and, following the election of Joe Biden, the United States – will further increase demand for the metals and minerals needed for electric vehicles and renewable energy technologies, and the major overhauls to national energy networks that these will require (Yu et al., 2021).

But the expansion of mining across Latin America has generated such intense social conflict that it is hard to envisage expansion on the massive scale that the energy transition will require – unless projects are imposed at gunpoint. While this does indeed happen in some cases, it is hardly a viable long-term strategy for either companies or governments, as the political and reputational risks are too high.

'They [the industry] themselves recognize that one of their biggest problems is obtaining this social licence; it's becoming increasingly difficult,' says Cesar Padilla, of the Latin American Observatory of Mining Conflicts (OCMAL). 'As mining expands, so does conflict, and so does rejection of the industry. And we believe this trend is only going to increase in the coming years.'

Knowledge is power

This rejection has occurred in radically different contexts and taken on many different forms: from direct action, sabotage, and armed confrontation; to peaceful protest, movement-building, and advocacy. It has even found expression in art, music, and street theatre.

From the communities' point of view, it is fortunate that this massive expansion of mining has occurred in the age of the internet and advanced communications technologies. Even in extremely remote and marginalized communities, mining opponents have used technology to their advantage, denouncing abuses committed by companies and governments and seeking international support. Resisting powerful transnational companies demands a global strategy, and today any community leader with a smartphone potentially has access to a global network of other mining-affected communities; journalists and media outlets; NGOs, aid agencies, and solidarity groups; churches and religious organizations; law firms; and academics.

The internet has also given communities access to a wealth of information on mining and its impacts; the ability to process and communicate this effectively is often key to successful resistance. This may involve critically assessing reports, press releases, and environmental impact studies by mining companies; researching similar scenarios elsewhere in the world; debating with politicians, civil servants, and the pro-mining media; and producing original communications material, amongst other tasks. This is no mean feat, and it's no coincidence that where levels of human development (and particularly education) are higher, it is often easier for mining opponents to build effective non-violent resistance.

Successful resistance can still occur in contexts of poverty, illiteracy, and state neglect, particularly if communities receive external support. But in general,

Photo 8.1 A MAB activist blocks the Minas-Vitória railway, which connects mining areas in Minas Gerais to port in Vitória, Espírito Santo / © Thais Gobbo 2019

it is more difficult. Without the same tools at their disposal, communities are more likely to resort to tactics such as roadblocks, which expose them to physical confrontations with the police, military, and private security guards, with sometimes lethal consequences. And if there is violence, the marginal position these communities occupy makes it easy for governments to blame them, smearing them as criminals, guerrillas, or terrorists. This reflects the historical class and racial prejudice which so often fuels violence associated with mining and other extractive industries (McNeish, 2018, p. 13).

Compare, for example, the experience of the anti-mining assemblies in Argentine towns and cities (see Chapter 1, Chapter 3, and Chapter 6), with that of campesinos in the Peruvian highlands (see Chapter 4) or indigenous groups in the Ecuadorian Amazon (see Chapter 7). This is not to say that the mining industry and the state in Argentina are models of good behaviour. Occasional outbreaks of violent repression occur, as in Chubut in late 2021; there are also cases of spying, threats, harassment, blacklisting of public employees, as well as academic and media censorship. Even so, the repression in Argentina falls far short of the military and paramilitary campaigns that have been waged on mining opponents elsewhere in the region.

'In the case of resistance to large-scale mining, the socio-environmental assemblies are very diverse. Lots of middle-class people participate, as well as indigenous people and campesinos,' says Lucrecia Wagner, from the National Scientific and Technical Research Council (CONICET) in Mendoza. 'This heterogeneity of actors has helped the assemblies to obtain greater social

Photo 8.2 Protestors stand together at the Rio Jáchal, San Juan, Argentina / © Oscar Martinez 2019

legitimacy. In other countries the level of violence is terrible, and yes there are killings, persecution, and human rights violations.'

'It's not that people aren't killed here, or that human rights are always respected, but here nobody is killed in mining conflicts. And if that were to happen, it would be a scandal.'

If not mining, then what?

In countries of long mining history in the region, particularly Peru and Chile, generations of policymakers – progressives as much as conservatives or neoliberals – have linked mining to national development, even to national identity. The popular Chilean expression '*El cobre es el sueldo de Chile*' ('copper is Chile's salary'), for example, is often associated with Salvador Allende and his nationalization of Chile's copper industry. Even today, the belief that resource extraction is the only path to a more prosperous and socially just future – as long as the state is in control, or at least gets its fair share of the profits – remains widespread.

It was manifest in the approach of recent Pink Tide governments, which sometimes explicitly linked popular social programmes and poverty reduction efforts to extractive industries. This made it possible for governments to cast critics of these industries not only as 'anti-development', but even as unpatriotic, elitists, enemies of the people (Boudewijn, 2020, p. 190), and indeed, it is harder to make the case against resource extraction when its

Photo 8.3 Dr Erik Jennings and his team at a Munduruku village / © Jorge Bodanzky 2019

rents are used to fund essential services which benefit the poorest and most vulnerable in society.

This deeply ingrained presupposition that mining and other extractive industries are a sure-fire route to prosperity also makes it harder to articulate alternatives. On top of denouncing the industry and supporting affected communities, mining opponents are constantly asked to provide their own solutions to complex and longstanding problems like poverty, unemployment, and economic stagnation, at both local and national level. This tends to boil down to the question 'If not mining, then what?' – to which there are two possible responses.

Firstly, that this is a loaded question, in that it presupposes ideas such as development and progress, modern Western concepts that may be alien or antithetical to traditional communities whose worldview is fundamentally distinct. Amongst some indigenous, Afro-descendant, and traditional campesino communities, there is hostility to the notion of development per se, a feeling that it is being imposed upon them from outside, or even that it is a kind of ideological trick used by wealthy countries and local elites to secure access to their natural and human resources (Machado Aráoz, 2012, p. 2).

'The debate was, well, if there's no large-scale mining, then what are we offering to the communities of the *meseta*?' explains Demián Morassi, referring to traditional communities in Chubut's *meseta central* which would have been impacted by the Navidad project (see Chapter 1). 'And our answer to that is that we don't have to offer them anything. The communities in the *meseta* know how they have to live; they've always lived there.'

This is not to romanticize traditional communities; many have long been neglected or even mistreated by their national states and experience deep-rooted social problems as a consequence. But it is to respect their right to self-determination, including the right to pursue traditional lifestyles without pressure to assimilate. That doesn't mean they reject modern education, healthcare, or technology; rather, they do reject – and indeed fear – projects which promise development but instead bring division and displacement, taking from them the little that they have while failing to provide them with the infrastructure and services they need.

The second possible response to the '… then what?' question is that even when communities accept development as a concept, it does not follow that they accept mining as the path towards it. Many mining opponents are not anti-capitalist; in fact, their suggested alternatives often involve a capitalist emphasis upon the modernization of methods and intensification of production. These alternatives vary from place to place, though the most frequently cited examples are tourism (including ecotourism) and agriculture. But they usually share some basic principles: firstly, that local people should take control of their own economy, via enterprises rooted in their own geography and culture; and secondly, that these enterprises should be sustainable, exploiting natural resources to create wealth and improve living standards, but without the negative long-term environmental impacts associated with mining (Boudewijn, 2020, p. 193).

For example, in Chile's Atacama Desert, the indigenous Atacameños have created a 'development plan', using the resources allocated to them under their agreement with the American lithium miner Albermarle (see Chapter 7), to promote tourism, agriculture, and other economic activities. On the one hand, this plan contains ideas on sustainability and shared use of resources rooted in their Andean indigenous worldview; on the other, it emphasizes improving production so they can be economically self-sufficient in the future. It is an expression of their dual identity: both as indigenous people, and as businesspeople.

'Our development plan is designed specifically to decouple us from mining, to no longer depend on it, to create a sustainable economy,' says Sergio Cubillos. 'But we can only work with the resources available to us. The state isn't putting resources into these communities, they don't say "Here, have this, do it." … So obviously the community has found it easier to deal with the private sector.'

This reflects a common concern amongst mining-affected communities, even if they don't articulate it in these terms: namely, the demand for a stronger and more responsive state, particularly at the local level. This goes far beyond just improved environmental protection and oversight of extractive activities. They want quality public services, particularly health and education; investment in local enterprise and access to credit for local farmers and businesspeople; and opportunities for young people, so they don't have to migrate to larger towns and cities. They want more of a say in what happens to their environment and greater power to improve their communities; in other words, they are demanding more democracy (Bebbington and Bury, 2013, p. 24).

'We're in favour of the preservation of our territory,' says Cubillos. 'We want to subsist as a community. We want sustainable development, with access to quality education and a healthcare system that meets our needs. We're not against the country continuing to progress and develop, but we think there needs to be a change in the vision of development ... Development means a sustainable, environmentally friendly economy – a system of life in harmony with our culture.'

'A paradigm shift'

One idea of great transformative potential, which may gain traction in the coming years, is that of the rights of nature: the notion that non-human animals and ecosystems are legal subjects with fundamental rights, in the same way as humans.

For many, the idea that an ecosystem like a river or forest could have fundamental rights enforceable in court may seem counterintuitive. Yet most of us broadly accept the idea that non-human animals are deserving of some rights, even if they fall short of those we afford to humans. We would also never question the entitlement of children, the elderly, and people who are severely disabled to rights, even though they depend on others to defend them – as would an ecosystem (Solón, 2022).

The rights of nature is an evolving legal approach, which departs from the conventional understanding of nature as a resource to be used for human benefit, towards one which recognizes that humanity has evolved as part of nature and depends upon healthy ecosystems for its survival (Harmony with Nature – United Nations, n.d.). It demands a shift in consciousness: from the anthropocentric, to the ecocentric.

'The rights of nature make it clear that the human species has a real and immediate dependence on nature, understood as the whole system of species and ecosystems of which we are also part,' explains David Fajardo Torres, an Ecuadorian law student and environmental activist with the groups Yasunidos and the People's Council for the Water of Cuenca. 'Anything that affects nature therefore means affecting humanity. In other words, if we don't first guarantee the rights of nature, it is impossible to guarantee human rights.'

In practice, the rights of nature oblige us to live within natural limits, establishing respect for the natural environment as a precondition for economic activity. This does not mean that nature must remain free from human intervention, but in theory it could preclude activities – like open-pit mining – which are so intensive they endanger nature's capacity to regenerate (Solón, 2022). This means sustainability is baked into the economic model, rather than being an afterthought, or an added extra to compensate for environmental harm.

'Amongst many other things, what the rights of nature signify is *limits*,' says Fajardo Torres. 'Limits even on the liberal conception of human rights ... For example, in the extractive industries, they often invoke the human right to work as a means of advancing with their operations. But the rights of nature allow us to put a limit on the right to work, as human rights are not absolute and above everything else. Of course people have the right to work ... But that right to work must be limited to activities which don't result in the elimination of ecosystems or biodiversity.'

Guided by indigenous Andean cosmovisions, Latin American countries have been the pioneers of rights of nature legislation. In 2008, Ecuador took the unprecedented step of enshrining the rights of nature in its national Constitution. This was followed by the 'Law of Mother Earth', passed in Bolivia in 2010, which recognized nature or Mother Earth as a legal subject with rights to be guaranteed by the state and wider society. Similar laws now exist in 37 countries worldwide, though to date no country has gone as far as Ecuador in affording nature constitutional protection (Acosta, 2022).

Sadly, the recent history of both Ecuador and Bolivia shows that without the political will to enforce such laws, their impact is limited. In the years since they recognized the rights of nature, both countries have doubled down on a model based on the intensive extraction of natural resources for export, particularly hydrocarbons and minerals, as well as intensive monoculture agribusiness. Those in both countries who have demanded that the rights of nature be upheld have faced repression, ranging from forced shutdowns of NGOs and restrictions on their access to funding, to state violence and criminalization of peaceful protest.

For example, in Ecuador, Rafael Correa – the president who oversaw the introduction of the 2008 Constitution – twice attempted to shut down the influential NGO Acción Ecológica. This occurred most recently in 2017, in retaliation for the NGO's support of Shuar communities in the Cordillera del Cóndor resisting the San Carlos-Panantza mining project (see Chapter 7). Though he was unsuccessful in this case, Correa did manage in 2013 to shut down the Fundación Pachamama, an NGO which works with indigenous communities in the Amazon which have resisted oil drilling, accusing it of political meddling and working to disrupt 'the internal security of the state' and 'public order' (Constante, 2013).

In Bolivia, Evo Morales and his vice-president Álvaro García Linera created a similarly hostile environment for NGOs, though many of these organizations had initially been supportive of the government. They were smeared as agents of foreign interests which were conspiring to destabilize the government and were subject to increased financial and bureaucratic pressures on their operations. The government even expelled the Danish NGO IBIS in late 2013 for alleged political interference, though the true reason may have been IBIS's support for an indigenous group which opposed Morales's plans to build a highway through the Tipnis, a national park and indigenous territory in the Amazon region (Achtenberg, 2015).

'The proposal to integrate the rights of nature into [national] constitutions ... is quite welcome,' says Marcos Orellana, an expert in international environmental law, currently the UN Special Rapporteur on toxics and human rights. 'It signifies a paradigm shift, where we abandon a purely anthropocentric approach, recognizing nature in its own right, and yet this can't be seen as a panacea. It doesn't act in isolation to safeguard the right to a healthy environment for present and future generations. That's because rights of nature can only be effective if accompanied by a vibrant civil society that defends them.'

The same could be said of human rights, though the concept has been around for much longer and is far more widely accepted. Human rights are frequently violated or only partially implemented; they constantly have to be defended, and further expansion of rights has to be fought for. But none of this makes human rights law redundant. On the contrary, it is precisely because such abuses and omissions continue to occur that it remains vital (Acosta, 2022).

Likewise, with about 200 species becoming extinct every day, scientists have warned that Earth's sixth mass extinction is already underway. The latest report by the UN's Intergovernmental Panel on Climate Change has warned that greenhouse gas emissions must peak before 2025 if there is to be a liveable future. We need an urgent revolution in our relationship with the natural world, and recognizing the rights of nature is an important first step.

'I think the rights of nature have enormous potential. Enormous and completely transformative for our societies,' says David Fajardo Torres. 'Especially in the context of climate change, biodiversity loss, and elimination of ecosystems in which we find ourselves, understanding the rights of nature as a limit on human societies is hugely important.'

***

In March 2022, Chile's constitutional assembly, the body responsible for writing Chile's new national constitution, voted in favour of including the following (Article 107):

> 'People and communities have an interdependent relationship with Nature and together form an inseparable whole.
>
> Nature has rights. The state and society have the duty to protect them and respect them.
>
> The state should adopt an ecologically responsible administration and promote environmental and scientific education via processes of ongoing training and study' (Convención Constitucional, 2022, p. 41).

Though the draft was rejected by voters in September 2022 (see Chapter 6), the fact that the rights of nature even featured in the document is significant.

It's not merely that Chile is a country of long mining tradition. It is a country in which the rights afforded to private property and private enterprise – going back to the radical neoliberal experiment imposed during the Pinochet dictatorship – are more extensive than anywhere else in Latin America, and indeed more than most other countries in the world. This has produced an economic model based on intensive extraction of commodities for export: not only mining, but also forestry, fishing, and agribusiness. Decades of this model have pushed ecosystems all over the country to the limit, resulting in pollution, desertification, soil degradation, and ever higher temperatures, leaving more than a million Chileans without a reliable water supply. Chile is also particularly vulnerable to climate change, which has fed into a vicious circle with these environmental impacts, generating a crisis which is becoming worse by the year.

The inclusion of the rights of nature in the draft is a sign that Chileans are increasingly realizing that business as usual is no longer an option. Whether Article 107 can be salvaged for the document that eventually becomes the country's new magna carta is unclear. But even if it ends up consigned to history, the environmental crisis which it aimed to address is going nowhere, and will increasingly influence the national political agenda. In Chile, as in the rest of the world, either we change – or change will be forced upon us.

Note

1.   All prices can be found at tradingeconomics.com

References

All references to web-based material were checked and still available in November 2022 unless otherwise stated.

All references are listed, with clickable links for your convenience, on the page for this chapter on the Heart of Our Earth website: <https://lab.org.uk/the-heart-of-our-earth/>

Achtenberg, E. (2015) 'What's Behind the Bolivian Government's Attack on NGOs?'. [online] NACLA. Available at: <https://nacla.org/blog/2015/09/03/what%27s-behind-bolivian-government%27s-attack-ngos>.

Acosta, A. (2022) 'Chile reconoce los derechos de la Naturaleza'. [online] *Clarín*. Available at: <https://www.clarin.com/opinion/chile-reconoce-derechos-naturaleza_0_4OxW7Gqz2P.html>.

Bebbington, A. and Bury, J. (2013) 'Political ecologies of the subsoil'. In: A. Bebbington and J. Bury, eds., *Subterranean Struggles. New dynamics of mining, oil, and gas in Latin America*. Austin, TX: University of Texas Press, pp. 1–26.

Boudewijn, I.A.M. (2020) 'Whose Development? How Women Living Near the Yanacocha Mine, Peru, Envision Potential Futures'. *Bulletin of Latin American Research*, 40(2), pp. 188–203.

Constante, S. (2013) 'Ecuador cierra una ONG que respaldaba la lucha antipetrolera en el Amazonas'. [online] *El País*. Available at: <https://elpais.com/internacional/2013/12/11/actualidad/1386772867_449366.html>.

Convención Constitucional (2022) *Borrador Nueva Constitución*. [online] Available at: <https://www.chileconvencion.cl/wp-content/uploads/2022/05/PROPUESTA-DE-BORRADOR-CONSTITUCIONAL-14.05.22-1-1.pdf>.

Harmony with Nature – United Nations (n.d.) 'Rights of Nature Law and Policy'. [online] Available at: <http://www.harmonywithnatureun.org/rightsOfNaturePolicies/>.

Machado Aráoz, H. (2012) 'Minería transnacional, conflictos socioterritoriales y nuevas dinámicas expropiatorias: el caso de Minera Alumbrera'. In: M. Svampa and M. Antonelli, eds., *Minería transnacional, narrativas del desarrollo y resistencias sociales*, 1st ed. [online] Buenos Aires: Biblos, pp. 181–204. Available at: <http://maristellasvampa.net/wp-content/uploads/2019/12/Miner%C3%ADa-transnacional.pdf>.

McNeish, J. (2018) 'Resource Extraction and Conflict in Latin America'. *Colombia Internacional*, (93), pp. 3–16.

Sandy, M. (n.d.) 'The Amazon Rain Forest Is Nearly Gone. We Went to the Front Lines to See If It Could Be Saved'. [online] *Time*. Available at: <https://time.com/amazon-rainforest-disappearing/>.

Solón, P. (2022) 'Chile aprueba los derechos de la Naturaleza'. [online] Fundación Solón. Available at: <https://fundacionsolon.org/2022/03/17/chile-aprueba-los-derechos-de-la-naturaleza/>.

Yu, A., Sappor, J., Nickels, L. and Cecil, R. (2021) 'Impact of COVID-19 pandemic on industrial metals markets - one year on'. [online] S&P Global Market Intelligence. Available at: <https://www.spglobal.com/marketintelligence/en/news-insights/research/impact-of-covid-19-pandemic-on-industrial-metals-markets-one-year-on>.

Index

Page numbers in *italics* refer to photos.

Ingram Content Group UK Ltd.
Milton Keynes UK
UKHW020748090423
419850UK00006B/74